Eugene Ritchie
Hill + Knowlton, Inc.
W9-BMA-127

# The Wall Street Journal

# Guide to

# The Metric System

*by*

Jerry C. Bishop

Published by Dow Jones Books
P.O. Box 300, Princeton, NJ 08540



Printed and bound in the United States of America
10 9 8 7 6 5 4 3 2 1

**Library of Congress Cataloging in Publication Data**

Bishop, Jerry.
The Wall Street journal guide to the metric system.

1. Metric system—Tables. 2. Metric system—United States. I. Title.
QC94.B54 389'.152'0212 78-4861
ISBN 0-87128-538-X

# TABLE OF CONTENTS

# Preface

## THE WALL STREET JOURNAL GUIDE TO THE METRIC SYSTEM

The metric system of measurement is about to become a fact of life for Americans. The requirements of advanced technology already have seduced scientists, engineers and physicians into daily use of metric measurements. U.S. industry has recognized that a change to metric measurements is a necessity if American enterprise is to keep its preeminent position in world commerce. And educators, statesmen, politicians, writers and businessmen are realizing that communication with the rest of the world requires adoption of the metric system.

This handbook is designed to help ease the transition to metric measurements in the home, the office and the school. It is designed primarily as a desktop reference that will enable the businessman to quickly convert customary measurements to metric units or metric measurements to customary units.

An indication of how this handbook can be best used might come from a brief look at its origins. It was conceived initially as an aid to Wall Street Journal reporters and editors covering developments in business and industry. Increasingly, news stories involve quantities of all sorts that are stated initially in metric units. This can prove quite irksome for an American newspaper reporter or copy editor, particularly when working hurriedly against a deadline. In order to restate the metric measurements in customary units that are more familiar to readers, the reporter or editor has to first rifle through reference books to find a conversion formula that tells, say, how many kilometers in a mile.

Once the conversion formula is found, a sometimes lengthy

multiplication or division has to be carried out either with paper and pencil or, more recently, with those ubiquitous little hand-held or desktop electronic calculators. The chance of making an error is quite high. Misplacing a decimal point can lead to major, sometimes ludicrous mistakes. Recently, for instance, the editor of a major newspaper chided his staff for printing a reference to "60 to 70 centimeter" diaphrams in a story on birth control. Such diaphrams would be an elephantine two feet in diameter. Obviously, the measurement should have been in millimeters instead of centimeters.

It was realized that the chances of making an error in converting metric measurements to customary units would be reduced if reporters and editors could quickly consult a table in which the conversions of various quantities had already been worked out. A glance at the table would quickly show, for instance, that 60 centimeters is equal to 23.6 inches and that 60 millimeters is equal to 2.36 inches.

It also became apparent that as the U.S. changed over to the metric system, newspapers would be faced with the additional problem of daily converting dozens, perhaps scores of quantities to metric units from customary units, converting pounds to kilograms, inches to centimeters, miles per hour to kilometers per hour, etc. In the weather report alone, temperatures, wind velocities, precipitation and barometric pressure all would have to be converted to metric units. Indeed, during the transition period, newspapers might find it necessary to give all measurements in both metric and customary units until readers became familiar with metric measurements. Here, again, a set of quick-reference tables in which the conversions from customary to metric units had already been worked out would be most useful and time-saving.

It was realized by the publisher that such problems of changing to the metric system won't be unique to the newspaper office. Americans in all walks of life from students, secretaries and shoppers to carpenters, accountants and corporate presidents will be faced almost daily with converting measurements to and from metric units. Hence, the decision to publish this handbook.

The major part of this handbook consists of tables, compiled with the help of a computer, converting measurements of all kinds from one system to the other. In total, almost 200 different units of measure are covered by the conversion tables. There are, for example, tables for 14 different units of length (ranging from inches and nautical miles to centimeters and kilometers) and 36 different units of volume (from fluid ounces and cords to liters and metric tablespoons).

Tables for conversions from customary to metric units are complimented with tables showing conversions from metric units to customary units. For instance, one table converts centimeters to inches while a second table immediately adjacent converts inches to centimeters.

With some exceptions, each table gives the conversions for 100 quantities. For example, the table converting centimeters to inches gives the computed conversions of one through 100 centimeters; a glance at the table will show, for instance, that 56 centimeters equals 22.047 256 inches. The user rarely will need to use a conversion carried to six decimal places; thus, in most cases, the user can round off the conversion to, in this case, 22.05 inches.

Conversion of quantities larger than 100 can be done quite easily by moving the decimal point the appropriate number of places. For instance, to find the conversion of 560 centimeters, look up the conversion of 56 centimeters (22.05 inches) and move the decimal point one place to the right, i.e. multiply by 10. Thus, 560 centimeters equals about 220.5 inches (or, to be more exact, 220.47 inches). Similarly, 5.6 centimeters equals 2.205 inches.

Since the metric system uses units that are related to each other by powers of 10, many of the tables can be used for conversions to more than one metric unit simply by moving the decimal point (i.e. multiplying by 10 or by 0.10). For example, to convert inches to millimeters use the table converting inches to centimeters. Since one centimeter equals 10 millimeters, the number of millimeters can be determined by moving the decimal point one place to the right. For instance, 45 inches equals 114.3 centimeters or 1143 millimeters. (See below for punctuation rules on metric numbers.) Similarly, since 100 centimeters equals one meter, the table can be used to convert inches to meters by moving the decimal point two places to the left; for example, 45 inches equals 1.143 meters.

A note of caution: Almost all of the conversions in the tables are carried out to six decimal places but the user shouldn't assume an accuracy to six decimal places. In some instances, rounding off of the conversion factor might have resulted in an inaccuracy in the last two decimal places, particularly in the larger quantities. Thus, if an accuracy to six decimal places is important, it's suggested the user re-compute the conversion using the formula in the section on Definitions and Conversion Factors.

The use of decimal arithmetic in the metric system discourages the use of common or vulgar fractions such as ⅛ or $^{1}/_{32}$. It is preferred that such fractions, instead, be converted to decimal fractions, i.e. 0.125 or 0.03125. A table showing the decimal equivalents of some of the more widely encountered common fractions immediately precedes the conversion tables.

This handbook isn't intended for use by scientists, engineers, and other specialists working in their own fields. It's assumed each discipline or field of expertise has its own guides and references for its unique problems of metric conversion. However, in the section on Definitions and Conversion Factors an attempt has been made to give the factors for converting many measuring units that are used only in highly specialized fields.

The first section of this handbook is a summary of the metric system, where it came from, how to use it and how to write it. This section is intended as a brief guide to using the metric system in everyday life. Those wishing to adopt the metric system for any particular industry or profession may desire more detailed descriptions of the system and its use. In such cases, it is suggested the reader determine and contact the appropriate trade or industry association or professional society charged with coordinating metric conversion in each industry or profession. Help in locating such coordinating groups can be had from the American National Metric Council, 1625 Massachusetts Ave., N.W., Washington, D.C., 20036.

Finally, the conversion to the metric system will affect almost every aspect of American life. To describe what these changes will be would require several volumes. A brief glimpse of some of the changes that the average American will encounter as he or she goes about daily living is provided in the second section of this handbook. Here the reader can find a description of a metric weather report, baking temperatures in degrees Celsius, or the sizes of metric wrenchs and screws.

# The Metric System

## A BRIEF HISTORY

Weights and measures were among the earliest tools invented by man. Primitive societies needed rudimentary measures for many tasks: constructing dwellings of an appropriate size and shape, fashioning clothing, or bartering food or raw materials.

Man understandably turned first to parts of his body and his natural surroundings for measuring instruments. Early Babylonian and Egyptian records and the Bible indicate that length was first measured with the forearm, hand, or finger and that time was measured by the periods of the sun, moon, and other heavenly bodies. When it was necessary to compare the capacities of containers such as gourds or clay or metal vessels, they were filled with plant seeds which were then counted to measure the volumes. When means for weighing were invented, seeds and stones served as standards. For instance, the "carat," still used as a unit for gems, was derived from the carob seed.

As societies evolved, weights and measures became more complex. The invention of numbering systems and the science of mathematics made it possible to create whole systems of weights and measures suited to trade and commerce, land division, taxation, or scientific research. For these more sophisticated uses it was necessary not only to weigh and measure more complex things—it was also necessary to do it accurately time after time and in different places. However, with limited international exchange of goods and communication of ideas, it is not surprising that different systems for the same purpose developed and became established in different parts of the world—even in different parts of a single continent. The measurement system commonly used in the United States today is nearly the same as that brought by the colonists from England.

These measures had their origins in a variety of cultures—Babylonian, Egyptian, Roman, Anglo-Saxon, and Norman French. The ancient "digit," "palm," "span," and "cubit" units evolved into the "inch," "foot," and "yard" through a complicated transformation not yet fully understood.

Roman contributions include the use of the number 12 as a base (our foot is divided into 12 inches) and words from which we derive many of our present weights and measures names. For example, the 12 divisions of the Roman "pes," or foot, were called *unciae*. Our words "inch" and "ounce" were both derived from that Latin word.

The "yard" as a measure of length can be traced back to the early Saxon kings. They wore a sash or girdle around the waist that could be removed and used as a convenient measuring device. Thus the word "yard" comes from the Saxon word "gird" meaning the circumference of a person's waist.

Standardization of the various units and their combinations into a loosely related system of weights and measures sometimes occurred in fascinating ways. Tradition holds that King Henry I decreed that the yard should be the distance from the tip of his nose to the end of his thumb. The length of a furlong (or furrow-long) was established by early Tudor rulers as 220 yards. This led Queen Elizabeth I to declare, in the 16th century, that henceforth the traditional Roman mile of 5,000 feet would be replaced by one of 5,280 feet, making the mile exactly 8 furlongs and providing a convenient relationship between two previously ill-related measures.

Thus, through royal edicts, England by the 18th century had achieved a greater degree of standardization than the continental countries. The English units were well suited to commerce and trade because they had been developed and refined to meet commercial needs. Through colonization and dominance of world commerce during the 17th, 18th and 19th centuries, the English system of weights and measures was spread to and established in many parts of the world, including the American colonies.

However, standards still differed to an extent undesirable for commerce among the 13 colonies. The need for greater uniformity led to clauses in the Articles of Confederation (ratified by the original colonies in 1781) and the Constitution of the United States (ratified in 1790) giving power to the Congress to fix uniform standards for weights and measures. Today, standards supplied to all the States by the National Bureau of Standards assure uniformity throughout the country.

The need for a single worldwide coordinated measurement system was recognized over 300 years ago. Gabriel Mouton, Vicar of

St. Paul in Lyons, proposed in 1670 a comprehensive decimal measurement system based on the length of one minute of arc in a great circle of the earth. In 1671 Jean Picard, a French astronomer, proposed the length of a pendulum beating seconds as the unit of length. (Such a pendulum would have been fairly easily reproducible, thus facilitating the widespread distribution of uniform standards.) Other proposals were made, but over a century elapsed before any action was taken.

In 1790, in the midst of the French Revolution, the National Assembly of France requested the French Academy of Sciences to "deduce an invariable standard for all the measures and all the weights." The Commission appointed by the Academy created a system that was, at once, simple and scientific. The unit of length was to be a portion of the earth's circumference. Measures for capacity (volume) and mass (weight) were to be derived from the unit of length, thus relating the basic units of the system to each other and to nature. Furthermore, the larger and smaller versions of each unit were to be created by multiplying or dividing the basic units by 10 or powers of 10 such as 100 or 1000, or 0.01 or 0.001. This feature provided a great convenience to users of the system, by eliminating the need for such calculations as dividing by 16 (to convert ounces to pounds) or by 12 (to convert inches to feet). Similar calculations in the metric system could be performed simply by shifting the decimal point. Thus the metric system is a "base-10" or "decimal" system.

The Commission assigned the name *metre* (which we also spell meter) to the unit of length. This name was derived from the Greek word *metron,* meaning "a measure." The physical standard representing the meter was to be constructed so that it would equal one ten-millionth of the distance from the north pole to the equator along the meridian of the earth running near Dunkirk in France and Barcelona in Spain.

The metric unit of mass, called the "gram," was defined as the mass of one cubic centimeter (a cube that is 1/100 of a meter on each side) of water at its temperature of maximum density. The cubic decimeter (a cube 1/10 of a meter on each side) was chosen as the unit of fluid capacity. This measure was given the name "liter."

Although the metric system was not accepted with enthusiasm at first, adoption by other nations occurred steadily after France made its use compulsory in 1840. The standardized character and decimal features of the metric system made it well suited to scientific and engineering work. Consequently, it is not surprising that the rapid spread of the system coincided with an age of rapid technological development. In the United States, by Act of Congress in 1866, it was made "lawful throughout the United States of America to employ the weights and measures of the metric system in all contracts, dealings or court proceedings."

By the late 1860's, even better metric standards were needed to keep pace with scientific advances. In 1875, an international treaty, the "Treaty of the Meter," set up well-defined metric standards for length and mass, and established permanent machinery to recommend and adopt further refinements in the metric system. This treaty, known as the Metric Convention, was signed by 17 countries, including the United States.

As a result of the Treaty, metric standards were constructed and distributed to each nation that ratified the Convention. Since 1893, the internationally agreed-to metric standards have served as the fundamental weights and measures standards of the United States.

By 1900 a total of 25 nations—including the major nations of Continental Europe and most of South America—had officially accepted the metric system. Today, with the exception of the United States and a few small countries, the entire world is using predominantly the metric system or is committed to such use. In 1971 the Secretary of Commerce, in transmitting to Congress the results of a three-year study authorized by the Metric Study Act of 1968, recommended that the U.S. change to predominant use of the metric system through a coordinated national program.

The International Bureau of Weights and Measures located at Sevres, France, serves as a permanent secretariat for the Metric Convention, coordinating the exchange of information about the use and refinement of the metric system. As measurement science develops more precise and easily reproducible ways of defining the measurement units, the General Conference on Weights and Measures—the diplomatic organization made up of adherents to the Convention—meets periodically to ratify improvements in the system and the standards.

In 1960, the General Conference adopted an extensive revision and simplification of the system. The name *Le Système International d'Unités* (International System of Units) with the international abbreviation SI, was adopted for this modernized metric system. Further improvements in and additions to SI were made by the General Conference in 1964, 1968, 1971 and 1975.

National Bureau of Standards
U.S. Department of Commerce
Special Publication 304 A

# WHAT THE METRIC SYSTEM IS AND HOW TO USE IT

The metric system is a system of measurement that is simpler and more logical than the customary or English system of measurement. A child who hasn't been indoctrinated in the customary measurement system can learn the metric system fairly quickly. The adult who has spent most of his or her life remembering that 12 inches equal a foot and 16 ounces equal a pound may approach the metric system with some trepidation because of the strangeness of such quantities as meters, kilograms and degrees Celsius. However, once the adult becomes familiar with only three or four basic metric units, the entire metric system falls into place and usually becomes the preferable measurement system.

**1. Base Units:** Any measurement system has to be built on at least three basic units—length, mass (commonly called weight) and time. In modern societies it also is necessary to have a basic unit of temperature and, as technology advances, basic units of electricity and light are needed. With the six basic units almost any conceivable measurement can be made, whether it be the speed of an automobile or the energy released by an atomic power plant.

The only real difference between the metric system and the customary system is in the six base units of measurement. Four of these base units—length, weight, time and temperature—are the ones most commonly used in everyday life. The unit of time—the second—is the same in both the metric and the customary systems. Thus, the key to learning to use the metric system in everyday life is in relearning only three base units—length, weight and temperature.

a. Length. In our customary system of measurement the foot is the base unit of length and all other units of length—the yard, the mile, the inch, etc.—can be defined in terms of feet or fraction of a foot. The length of a foot has been agreed upon only by custom.

In the metric system, the meter is the base unit of length and all other units of length, from the tiny millimeter to the kilometer, are defined in terms of the meter. Originally, the meter was defined as a certain fraction of the circumference of the earth, a standard which could be measured anytime anywhere. It has since been redefined and in our present high-technology society it is defined as being equal to the length of a certain number of light waves.

Because the length of the meter is based on an unvarying object in nature, it doesn't come out to be an exact multiple of the customary foot. As it turns out, the foot is now defined as equal to a certain fraction of the meter. Specifically, the foot is 0.304 800 part of a meter. This, in turn, makes the meter equal to 3.280 840 feet.

As a rough rule of thumb, the meter can be thought of as being about 39 inches or slightly longer than a yard.

b. Weight. In the customary system the base unit of weight is the avoirdupois pound of 16 ounces, also a quantity defined by custom over the centuries. In the metric system, the base unit of weight is the kilogram. The kilogram is defined by international agreement as being equal to the mass of a precision-made cylinder of platinum-iridium alloy that is stored in a vault in Sevres, France. The U.S., as well as most other nations, has a duplicate of this cylinder against which scales and other instruments for measuring weight and mass can be calibrated.

The pound is now defined as being 0.453 592 370 kilogram. This makes the kilogram equal to 2.204 622 pounds.

In everyday life, the kilogram can be thought of as being roughly equal to 2.2 pounds while the pound can be thought of as, roughly, slightly less than half a kilogram.

c. Temperature. In the customary system the unit of temperature is the degree Fahrenheit. Unlike the units of length and mass, the degree Fahrenheit isn't based on custom but rather on a phenomenon in nature. Gabriel D. Fahrenheit in Holland in 1714 devised a temperature scale that used the freezing point of a certain solution of salt water as zero and the blood temperature of a healthy man as 96 (an error of 2.6 degrees it was discovered later). This scale makes the freezing point of pure water about 32 degrees and the boiling point of water about 212 degrees.

The commonly-used temperature unit of the metric system is the degree Celsius, named after Anders Celsius, an 18th Century Swedish astonomer who invented the centigrade scale. It is based on setting 0 degrees at the freezing point of water and 100 degrees at the boiling point of water. (In everyday life, temperatures are given in degrees Celsius although because of the needs of scientists and engineers the metric base temperature unit is kelvin unit. A temperature change of one degree is the same in the Celsius and kelvin scales but the kelvin scale begins with absolute zero or the absolute absence of all heat, a point so cold—minus 459.67 degrees Fahrenheit—as to be irrelevant for everyday use.)

There isn't any easy way to make rough conversions to and from Fahrenheit and Celsius temperatures. Exact conversions can be made as follows:

Take the temperature in degrees Fahrenheit, subtract 32, and multiply by 5/9 to obtain the temperature in degrees Celsius. (Often it is easier to multiply by 0.556 rather than 5/9 but the conversion won't be exact.)

Take the temperature in degrees Celsius, multiply by 9/5 and add 32 to obtain the temperature in degrees Fahrenheit. (A slightly easier but less exact method is to add 17.8 and then multiply by 1.8 to obtain Fahrenheit.)

In summary, there are six units in a modern measurement system with which all measurements are made. A comparison of the base units of the customary and metric system is as follows:

| Base Unit | Metric | Customary |
|---|---|---|
| Length | meter | foot |
| Weight | kilogram | pound |
| Time | second | second |
| Temperature | degree Celsius | degree Fahrenheit |
| Electric current | ampere | ampere |
| Luminous intensity | candela | candela |

(In 1967, the mole was added as the base unit for the amount of a substance, a quantity used by chemists and other scientists.)

Exact definitions and conversion formulas for the base units can be found in the section on definitions and conversion factors.

**2. Derived units:** A large number of other units of measurement are needed in a modern society such as units of velocity, acceleration and pressure. These are derived from two or more of the base units. Velocity, for example, is derived from taking a certain length (or distance) and dividing it by the time it takes to travel that length.

Since some of the base units in the metric system are different than the base units in the customary system, derived units also are different. For example, the metric unit of velocity is meters per second (often converted to kilometers per hour) while the velocity unit in the customary system is feet per second (often converted to miles per hour).

In the metric system some of these derived units are given specific names, often to honor a famous scientist. For example, the unit of force, comparable to the "pound-force" in the customary system, is called the newton in the metric system. (The newton is defined as the force it takes to accelerate one kilogram of mass at the rate of one meter per second each second.) There are several other so-named derived units such as the watt (power) and the joule (energy).

A few examples of derived units are as follows:

| Derived Unit | Customary | Metric |
|---|---|---|
| Area | square foot | square meter |
| Volume | cubic foot | cubic meter |
| Velocity | feet/second | meters/second |
| Force | pound-force | newton |
| Pressure | lb/square foot | pascal (newton/square meter) |
| Work, energy | foot-pound | joule (newton-meter) |

Definitions of these and other derived units can be found in the section on definitions and conversion factors. Tables converting various quantities of metric units to and from customary units can be found in the section of conversion tables.

# METRIC ARITHMETIC

A major appeal of the metric system is that it uses the arithmetic of decimal numbers. This is an arithmetic based on the number 10 and powers of 10.

Decimal arithmetic should come quite easy to Americans since their currency is based on the decimal system. The base unit of our decimal currency is the dollar, written $1.00. This base unit is divided into hundredths (0.1 × 0.1), written as $0.01. Any quantity of money can be stated in tenths and hundredths of a dollar and in tens, hundreds, thousands, etc. dollars by using decimal numbers, i.e. $10.25 denotes ten times one dollar plus 25 hundredths of a dollar. As a matter of convenience we call the one-hundredth of a dollar a cent, from the same Latin word for hundred that is the root word for century and centipede. We can, if we desire, write 25¢ instead of $0.25 in which case we've converted dollars to cents by multiplying the dollars amount by 100.

Metric measurements are written and calculated in the same way as U.S. currency. For example, the base unit of length is the meter, written 1.00 m (m being the symbol for meter). The meter, like the dollar, is divided into hundredths, written as 0.01 m. For convenience, the hundredth of a meter is given the name, centimeter. Thus, 25 hundredths of a meter can be written as 0.25 m or as 25 cm. In the later case, meters are converted to centimeters by multiplying by 100 just as dollars are converted to cents.

All metric measurements can be stated in the base units (or in many cased the derived units) as in 1 000 m, for instance. However, it may be somewhat tedious and inconvenient to write very large or very small quantities only in base metric units as, for instance, writing 1 000 000 meters. Therefore, the metric system has units denoted by prefixes for very large and very small quantities.

It is important to remember that our customary system has specially-named units for quantities that are larger or smaller than the base units. The base unit of length is the foot but in dealing with large distances we convert feet to miles (at the ratio of 5,280 feet to the mile) and in dealing with small distances we convert feet to inches (at 12 inches to the foot).

In the metric system the larger units are always the base unit times some power of 10 while the smaller units are always the base unit divided by some power of 10. Just how much the base unit is multiplied or divided by is denoted by a prefix. For instance, the prefix "centi-", for one hundredth, denotes the unit is 0.01 of the base unit. Thus, for very short lengths one can deal in centimeters instead of meters, as, for example, 2.5 centimeters instead of 0.025 meters. Similarly for very long lengths or distances, one can deal in kilometers, the prefix "kilo-" denoting 1 000, as, for example, 250 kilometers instead of 250 000 meters. (A complete list of the prefixes can be found in the section "How to Write Metrics".)

Note how much easier it is to convert metric measurements than to convert customary measurements. In the customary system we have to convert miles to feet by multiplying by 5,280 whereas in the metric system we can convert kilometers to meters merely by multiplying by 1 000. Since we use decimal arithmetic with metric measurements, multiplying by 1 000 merely involves moving the decimal point three places to the right (12.345 kilometers equals 12 345.0 meters). Similarly, in weights in the customary system we have to convert pounds to ounces by multiplying by 16 but in the metric system we can convert kilograms to grams by multiplying by 1 000 (thus, 1.234 kilograms equals 1 234 grams).

In decimal numbers, of course, all whole numbers are written to left of the decimal point and all fractions are written to the right of the decimal point (that is, 0.1 is a tenth, 0.01 a hundredth, 0.001 a thousandth, etc.).

In the metric system common or vulgar fractions such as ⅓ or

⅛ are seldom used, (although ½ and ¼ probably will continue to be used out of habit). Instead of writing, say, ⅛, it is preferable to use the decimal equivalent, 0.125. A list of the decimal equivalents of the more widely used common fractions can be found at the beginning of the section of conversion tables. This makes arithmetic in metric measurements far easier than in customary measurements. It is considerably easier to multiply or divide 1.125 by 2.25 than it is to multiply or divide 1⅛ by 2¼, for example. Indeed, the little hand-held electronic calculators that are in such wide use can deal only in decimal fractions, a fact which, more than anything else, may spur acceptance of the metric system.

# WRITING METRIC

Unlike the haphazard "customary" system of weights and measures, where the rules of writing style often seem without logic (lb. for pound, for example), the International System of metric measurements uses a well-planned and specific system of rules of spelling, punctuation and abbreviation. The rules of style are designed to avoid confusion between writer and reader and should be followed carefully. For example, a capital M is the symbol for the prefix "mega-", denoting a million or 10 to the sixth power. A lower case m, however, is the symbol for the prefix "milli-", denoting one-thousandth or 10 to the minus three power. Obviously, it is extremely important that a writer describing large amounts of electric power, say, use the symbol MW for megawatts rather the mW (with a lower case m) which is the symbol for milliwatt, an extremely small amount of power. The lower case m also happens to be the abbreviation for meter but as one becomes familiar with the metric system he or she will find it is immediately clear from the context of the symbol whether m stands for meter or milli-.

In some instances, SI rules call for the use of symbols that cannot be written with a standard typewriter or typesetting machine. The Greek letter "μ," for instance, is the symbol for the prefix, micro-, denoting millionths. Greek letters, however, are available only on typewriters and typesetters used for engineering and scientific publications. Most writers and publishers, therefore, will have to use alternate, less desirable symbols. The symbol "my," for instance, can be substituted for the Greek letter "μ."

The following style rules are designed for use with the standard English-language typewriter and are those recommended for newspapers and other non-technical publications.

Computer users are referred to *Dun & Bradstreet's Guide to Metric Transition for Managers* by Robert C. Sellers (Thomas Y. Crowell Co., New York, 1975) for the appropriate writing style used for computer printouts. Scientists and engineers are referred to the *Metric Practice Guide* published by the American Society of Testing and Materials (1916 Race St., Philadelphia, PA., 19103).

1. **Prefixes** It is much easier for both the writer and the reader to use certain prefixes instead of zeroes to denote multiples and submultiples of various quantities. It is easier, for example, to give a distance as 100 kilometers rather than 100 000 meters or to write 10 millimeters instead of 0.010 meter.

| Prefix | Power of 10 | Symbol | U.S. Name |
|---|---|---|---|
| deka- | 1 | da | ten |
| hecto- | 2 | h | hundred |
| kilo- | 3 | k | thousand |
| mega- | 6 | M | million |
| giga- | 9 | G | billion |
| tera- | 12 | T | trillion |
| peta- | 15 | P | quadrillion |
| exa- | 18 | E | quintillion |
| deci- | −1 | d | tenth |
| centi- | −2 | c | hundredth |
| milli- | −3 | m | thousandth |
| micro- | −6 | my | millionth |
| nano- | −9 | n | billionth |
| pico- | −12 | p | trillionth |
| femto- | −15 | f | quadrillionth |
| atto- | −18 | a | quintillionth |

(Note that the symbols of all prefixes are lower case except for the largest quantities, i.e., mega-, giga-, tera-, peta-, and exa-.)

Only one prefix can be used at a time; compound prefixes are incorrect. Example: write nanometer instead of millimicrometer.

2. **Symbols** Below are accepted symbols for SI units. Generally, those symbols for units named after historic persons are capitalized even though the unit, itself, is lower case.

| Unit | Symbol |
|---|---|
| ampere | A |
| candela | cd |
| Celsius degree | °C |
| coulomb | C |
| farad | F |
| gram | g |
| henry | H |
| hertz | Hz |
| joule | J |
| kelvin | K |
| liter | L* |
| lumen | lm |
| lux | lx |
| meter | m |
| mole | mol |
| newton | N |
| ohm | ohm |
| pascal | Pa |
| radian | rad |
| second | s |
| siemens | S |
| steradian | sr |
| tesla | T |
| ton | t* |
| volt | V |
| watt | W |
| weber | Wb |

The **plural** is never used with metric symbols since the letter s is the symbol for second. Write 4.5 km **not** 4.5 kms which might be interpreted as kilometer-seconds, an entirely different measurement.

Metric symbols are **never italicized.**

A single or half space should separate the quantity and the symbol. Example, 4.5 m, not 4.5m.

Symbols for the prefix and the unit should be typed or printed without spacing. Example: 4.5 km, not 4.5 k m.

*Because a lower case 1 can be confused with the number 1, it is recommended liter be spelled out or, the symbol L be used. It is preferable to write out ton instead of using a lower case t.

3. **Punctuation** The **period** is used as a decimal point in the U.S.

**Do not** use a period after metric symbols except at the end of a sentence; these are symbols, not abbreviations. Example: write 4.5 m (without the period).

The **comma** isn't used in writing metric measurements in North America. In writing **lengthy numbers,** groups of three digits are separated by a space instead of a comma. Example: 132 456 000. This avoids mistaking a comma for a decimal point and prevents confusion with the practice in some countries of using a comma instead of a period as a decimal marker.

The **hyphen** isn't used with metric symbols. However, in writing the compound name of a metric unit that is the product of multiplication a hyphen may be used instead of a space to separate the multiplicand and the multiplier. Example: meter kilogram can be written, meter-kilogram.

The **solidus** or **"slant-mark"** (/) denotes division, that is, it separates the dividend and the divisor, in metric symbols. Example: kilometers per hour would be km/h (not, incidentally, kph). Use only one solidus in any combination of symbols: $km/s^2$, not km/s/s.

4. **Fractions** Vulgar or common fractions such as ½ or ¼ may be used in the metric system but it is preferable to use decimal fractions, such as 0.5 or 0.25. A whole number or a zero should always precede the decimal point in writing fractions. A table showing the decimal equivalents of common fractions appears at the beginning of the section of conversion tables.

5. **Squared and cubed** measurements should be designated by the appropriate exponent written as a superscript. Example: square kilometer is written $km^2$ while cubic centimeter is written $cm^3$ (the use of cc for cubic centimeter is incorrect as is sq.km. for square kilometer). Generally, it is preferable to use symbols for units of area and volume instead of writing them out, that is, use $km^2$ instead of square kilometers.

# Metric Measurements in Everyday Life

## WEATHER

For most Americans, the first jarring encounter with the metric system comes when the television or radio weatherman gives the day's temperatures in degrees Celsius as well as degrees Fahrenheit. As the transition to the metric system takes place, other weather measurements also will be given in metric units: snowfall will be centimeters and rainfall will be in millimeters instead of inches, wind speeds in kilometers per hour instead of miles per hour and barometric pressures in kilopascals instead of inches of mercury.

Few areas of everyday life have triggered more debate on how to go about changing to the metric system than the weather forecasts. Many metric system proponents argue that weather statistics should be given solely in metric units. The initial confusion, they say, will be short-lived as people learn by experience that, say, 20°C is a comfortable room temperature while 40°C is a hot summer day. This is the basis for the Canadian government's "Think Metric" policy under which weather reports and forecasts, as well as all other measurements, are stated only in metric units.

In the U.S., however, radio, television and newspapers appear reluctant to use only metric units in the weather reports. Apparently worried that they'll stir the enmity of listeners and readers by using the unfamiliar metric units, the stations and newspapers often give the reports in both metric and customary units. As the metric units become more familiar they will gradually drop the customary units.

Complete conversions of temperatures, presures, velocities and precipitation measurements to and from metric and customary units can be found in the section of conversion tables.

For those who hear a weather report using only metric units, the following "benchmarks" might be helpful.

**Temperature** in degrees Celsius:

| | |
|---|---|
| 100 | water boils |
| 38-41 | heat wave |
| 29-37 | warm to hot summer day |
| 21-28 | pleasant spring or autumn day |
| 20 | room temperature |
| 10-19 | cool |
| 1-9 | cold |
| 0 | water freezes |
| −1 to −10 | antifreeze in the auto |
| −11 to −20 | very cold for New York City |
| −21 to −30 | very cold for Minnesota |

**Snowfall** in centimeters:

| | |
|---|---|
| 1-3 | Light snowfall |
| 3-9 | Moderate snowfall |
| 10-14 | Snowfall requiring shoveling of walks and plowing of streets. |
| 15 plus | Heavy snowstorm |
| 25-30 | Extremely heavy overnight snowfall for New York or Chicago. |
| 50-60 | Average annual precipitation for Southwest U.S. |
| 70-100 | Average annual precipitation for middle U.S. |

**Wind velocity,** in kilometers per hour:

| | |
|---|---|
| 1-20 | light wind, extends a flag, rustles leaves |
| 21-40 | moderate wind, raises dust and loose paper |
| 41-60 | strong wind, breaks umbrellas, drifts snow |
| 61-90 | gale winds, damages trees and roofs |
| 91-115 | storm winds, downs trees, damages buildings |
| 115 plus | hurricane |

**Barometric pressure,** in kilopascals:

| | |
|---|---|
| 98 to 103 | is the normal range of atmospheric pressure. |

| | |
|---|---|
| 95-98 | Low pressure, stormy, rain. |
| 98.5-101.5 | medium pressure, weather changing |
| 102-105 | High pressure, fair and dry. |

# METRICS IN THE SUPERMARKET

Generally, the supermarket shopper should have little difficulty changing to metric units. The change to metric units should make it far easier for the shopper to "comparison shop," that is, to determine the price per unit of weight or volume, to quickly estimate which brand or package of a food offers the best bargain. The biggest problem lies not with the shopper learning the metric system but with getting the food processors to change their labeling and to standardize and make more uniform the plethora of different sized cans, bottles and boxes. When the packaging becomes more uniform—and the transition to the metric system is an opportunity to carry out this reform—the use of grams and kilograms, liters and milliliters should make shopping considerably easier than the present confusing practice of using ounces, pounds, fluid ounces, cups, pints, quarts, pecks, bushels, etc.

Exact conversions of ounces, pounds, bushels, quarts, etc., to and from metric units can be found in the section of conversion tables elsewhere in this book.

Following, however, are some general guidelines and hints that may prove helpful to the "metric shopper".

**Meats, fresh fruits and vegetables** which are usually sold by weight will be measured in grams or kilograms. The weights will be stated only in one unit; that is, it will be 1.45 kilograms or 1450 grams (never 1 kilogram 450 grams). This should pose few problems as long as the shopper remembers that one kilogram is about 2.2 pounds; the shopper accustomed to buying one pound quantities should think in terms of 500 grams while quarter-pound quantities should be mentally changed to 250 grams.

A typical **individual serving** of boned meat is about 85 to 100 grams (roughly three to 3½ ounces). A typical serving of fruits or vegetables is a metric cup (250 milliliters).

**Dairy products** that normally are sold by the pint and quart will be sold by the liter whereas those normally sold by the dry ounce will be sold by grams. A **liter of milk** is slightly larger than a quart (specifically a liter equals about 1.057 quart); two liters is about a half-gallon while four liters is a fraction more than a gallon. A half-liter or 500 milliliters is about the same as a pint.

Typical individual servings of dairy products are 250 milliliters of milk (about the same as an 8 oz. glass), 85 to 100 grams of cheese (about three 1 oz. slices), 120 milliliters of cottage cheese (about half a cup) or 250 milliliters of plain yogurt (about a cup).

**Boxed, canned and bottled** foods already are being dually labeled by some food processors with the contents being stated in grams for foods sold by weight and milliliters for those sold by volume. To completely change over to the metric system, however, food processors should change their package sizes so that each package will be in more or less even, easy-to-remember metric dimensions or quantities.

Just what changes U.S. food processors will make in package sizes isn't yet settled. However, a look at some changes made in Britain might provide some suggestion of package changes.

**Breakfast cereals:** boxes containing 375 g and 500 g replaced the traditional 12 ounce and 16 ounce cereal boxes.

**Butter, margarine and cooking oils** in Britain are sold in 50 g, 125 g, 250 g, 500 g, 1 000 g (1 kilogram) sizes generally. Large sizes, up to 4 kg, come in multiples of 500 g, while quantities of 4 kg to 10 kg are packaged in multiples of 1 kg.

**Dried fruit** is packaged in 125 g, 250 g, 375 g, 500 g and 1 kg sizes, with large sizes in 1 kg multiples up to 10 kg.

**Crackers and cookies** are packaged in 100 g, 125 g, 150 g, 200 g, 250 g, and 300 g weights with larger sizes coming in multiples of 100 g up to 5 kg.

**Sugar and flour** will be packaged in 1 kg, 2 kg, 4 kg, 10 kg and 40 kg sacks instead of the traditional 2, 5, 10, 25 and 100 pound sacks. A 4 kg sack of sugar will replace the 10 pound bag but note that 4 kg is equal to 8.8 pounds and should be priced 12% lower than the 10 pound bag.

**Liquor** bottles already are being changed to metric sizes in the U.S. with the standard "fifths" and quarts being changed to even liter sizes. The imbiber will find the following new bottles:

| metric | customary |
|---|---|
| 50 ml (1.7 ounces) | miniature (1.7 ounces) |
| 200 ml (6.8 ounces) | half-pint (8 ounces) |
| 500 ml (16.9 ounces) | pint (16 ounces) |
| 750 ml (25.4 ounces) | fifth (25.6 ounces) |
| 1 liter (33.8 ounces) | quart (32 ounces) |
| 1.75 liter (59.2 ounces) | half-gallon (64 ounces) |

# METRICS IN THE KITCHEN

The American cook can rest easy. All the old recipes and cookbooks will remain valid and useable even after the change to metric measurements is completed. When a recipe calls for, say, one cup of flour and one teaspoon of salt, it will make little difference whether the cook uses the old customary measuring cup and teaspoon or the new metric cup and teaspoon. Although the new metric measuring utensils will be slightly different in size from the customary utensils, the difference is too small to affect the outcome of most recipes used in cooking for the family. It is only when one is cooking very large quantities, as for a church dinner or hotel or an army, that the differences between customary and metric measurements become significant.

The reason why there should be little difficulty in continued use of older recipes is evident to almost anyone who has watched different cooks at work in the kitchen. One cook, for instance, might always use a slightly heaping teaspoon while another will always measure a level teaspoon yet the end result for both cooks will be the same. Studies have shown that the quantities used by different cooks preparing the same recipe will vary as much as 5%. Recipe writers long have taken this variation between individual cooks into account. As a result, the quantities stated in most recipes are approximate amounts rather than exact amounts.

The difference in sizes between the new metric kitchen utensils and the old customary utensils is smaller than the variations permitted in most recipes. Thus, use of a metric measuring cup and spoon with old recipes shouldn't affect the quality of one's cooking.

As the changeover to the metric system gets underway, however, the cook may find it easier to discard the old customary measuring cups and spoons and to begin using the new metric measuring utensils. The reason is that recipes may soon be stated in metric units rather than customary units. A recipe, for example, may call for 250 milliliters of heavy cream instead of one-half pint or eight ounces. Obviously, if one had a metric measuring cup graduated in milliliters it would be quite simple to follow such a recipe.

The dimensions of cooking pots and pans will be changed to centimeters from inches. The size change will be hardly noticeable, however, since its main purpose will be to make the dimensions come out in whole centimeters rather than fractions. For example, a 10-inch pie plate is 25.4 centimeters. In order to avoid such fractional dimension, the nearest metric equivalent to the 10-inch pie plate is named the 25-centimeter plate. The problem the cook may encounter is learning to ask the store clerk for a 25-centimeter pie plate and learning which plate to reach for when the recipe calls for a 25-centimeter pie plate.

The biggest problem the cook may encounter in switching to metric measurements might be in cooking temperatures. The temperatures will be stated in degrees Celsius instead of degrees Fahrenheit. Thus, a recipe may say to bake two hours at 165° C instead of 325° F.

Following are some of the details of some of the changes that will be made in the kitchen as the country switches over to the metric system.

**1. Measuring cups and spoons:** A new "metric cup" will replace the customary cup. The metric cup will be 250 mL in liquid volume or 250 cubic centimeters in dry volume. The metric cup is slightly larger than the customary cup which is 237 mL; the metric cup, in other words, is about a tablespoon larger than the customary cup.

A metric measuring cup will be marked along the sides in gradations of milliliters. For example, every 25 mL might be marked or, perhaps, every 10 mL, making it possible to measure 50 mL, 100 ml, 125 ml, etc. It also will be marked at the quarter, one-third, one-half, two-thirds, and three-quarters levels.

The customary U.S. teaspoon, unfortunately, is equal to an unwieldy multiple of a milliliter, 4.9289 mL to be exact. Obviously, if measurements in the kitchen are going to be in milliliters it makes sense if the teaspoon were a more easily used multiple of a milliliter. For this reason, the metric teaspoon is 5.0 mL. This way 50 teaspoons equal exactly one metric cup (250 mL).

A similar situation exists with the tablespoon. The customary tablespoon equals 14.786 800 mL. To make it easier to use, the metric tablespoon is defined as exactly 15 mL, making it fractionally larger than the customary tablespoon.

A set of metric measuring spoons probably will consist of the usual four spoons: a tablespoon (15 mL), a teaspoon (5 mL), one-half teaspoon (2.5 mL) and one-fourth teaspoon (1.25 mL).

There have been suggestions that a new basic metric spoon be introduced. This basic spoon would be 10 mL and would simplify considerably the use of metric measurements in the kitchen; for instance, any quantity could be measured with only two basic spoons, a 10 mL spoon and a 1 mL spoon. If the people who write recipes begin stating quantities in milliliters such a basic 10 mL spoon may come into use.

In summary, the "metric cook" can discard the old, somewhat irrational measures of 16 tablespoons to the cup, four cups to the quart, etc. and use the following equivalent measures:

| | Metric Utensil | Quantity (ml) | | Nearest U.S. Customary Measure | |
|---|---|---|---|---|---|
| ¼ | teaspoon | 1.25 | ¼ | teaspoon | (1.23 mL) |
| ½ | " | 2.5 | ½ | " | (2.46 mL) |
| 1 | " | 5.0 | 1 | " | (4.93 mL) |
| 1 | tablespoon | 15.0 | 1 | tablespoon | (14.8 mL) |
| ¼ | cup | 62.5 | ¼ | cup | (59.15 mL) |
| ⅓ | " | 83.3 | ⅓ | " | (78.9 mL) |
| ½ | " | 125.0 | ½ | " | (118.3 mL) |
| 1 | " | 250.0 | 1 | " | (236.6 mL) |
| ½ | liter | 500.0 | 1 | pint | (473.2 mL) |
| 1 | " | 1 000.0 | 1 | quart | (946 mL) |
| 4 | " | 4 000.0 | 1 | gallon | (3 785 mL) |

**2. Weights:** In recipes where ingredients are specified by weight, such as six ounces or 170 g, exact conversions from ounces to grams and grams to ounces can be found in the conversion tables section of this book. As a rough guide, however, the following might be useful:

| 1 metric cup of | equals | grams |
|---|---|---|
| butter | " | 236 |
| salad oil | " | 222 |
| flour, sifted | " | 121 |
| sugar, white or brown | " | 211 |
| cocoa | " | 118 |

**3. Large Quantities:** For church supers, banquets or other affairs, where one might want to triple or quadruple the amounts specified in ordinary recipies, the differences between customary cups and spoons and metric cups and spoons might be sufficient to affect the outcome of a dish. This may also be true when canning a large volume of a food.

A set of customary spoons and cups can be used to measure milliliters and metric cups as follows:

| | | | | | | | |
|---|---|---|---|---|---|---|---|
| | ½ | customary | teaspoon | equals | 2.5 | milliliters | |
| | 1 | " | " | " | 4.9 | " | |
| | 2 | " | " | " | 9.9 | " | |
| | 1 | " | tablespoon | " | 14.8 | " | |
| | 2 | " | " | " | 29.6 | " | |
| | 3 | " | " | " | 44.4 | " | |
| (¼ cup) | 4 | " | " | " | 59.1 | " | |
| | 5 | " | " | " | 73.9 | " | |
| | 6 | " | " | " | 88.7 | " | |
| | 7 | " | " | " | 103.5 | " | |
| (½ cup) | 8 | " | " | " | 118.3 | " | |
| | 9 | " | " | " | 133.1 | " | |
| | 10 | " | " | " | 147.9 | " | |
| | 11 | " | " | " | 162.7 | " | |
| (¾ cup) | 12 | " | " | " | 177.4 | " | |
| | 13 | " | " | " | 192.2 | " | |
| | 14 | " | " | " | 207.0 | " | |
| | 15 | " | " | " | 221.8 | " | |
| (1 cup) | 16 | " | " | " | 236.6 | " | or 0.9 metric cup |
| | 2 | customary | cups | equals | 473.2 | " | " 1.9 " " |
| | 3 | " | " | " | 709.8 | " | " 2.8 " " |
| | 4 | " | " | " | 946.4 | " | " 3.8 " " |
| | 5 | " | " | " | 1 182.9 | " | " 4.7 " " |
| | 6 | " | " | " | 1 419.5 | " | " 5.7 " " |
| | 7 | " | " | " | 1 656.1 | " | " 6.6 " " |
| | 8 | " | " | " | 1 892.7 | " | " 7.6 " " |
| | 9 | " | " | " | 2 129.3 | " | " 8.5 " " |
| | 10 | " | " | " | 2 365.9 | " | " 9.5 " " |

Milliliters can be converted to customary spoons and cups as follows:

| | | | | | | | | |
|---|---|---|---|---|---|---|---|---|
| 10 | mL | equals | 2 | tspn | | | | |
| 50 | " | " | 10 | " | or | 3⅓ | tbsp | |
| 100 | " | " | 20¼ | " | " | 6¾ | " | |
| 125 | " | " | | | | 8½ | " | |
| 150 | " | " | | | | 10 | " | |
| 175 | " | " | | | | 11¾ | " | |
| 200 | " | " | | | | 13½ | " | |
| 237 | " | " | | | | 16 | " | or 1 cup |
| 250 | " | (or 1 metric cup) equals | | | | 1 | " | plus 1 cup |
| 500 | " | (or 2 metric cups) " | | | | 2 | " | plus 2 cups |
| 750 | " | (or 3 metric cups) " | | | | 3 | " | plus 3 cups |
| 1 000 | " | (or 1 liter) " | | | | | | 4¼ cups |

**4. Pots and pans:** The principal change in "going metric" will be that the dimensions of pots and pans will be stated centimeters rather than inches. The pots and pans could remain the same size except this poses an inconvenience in that the dimensions would have to be given in fractions of a centimeter. It is more convenient for manufacturers and suppliers to round off the dimensions to the nearest full centimeter.

The following metric dimensions for pots and pans were suggested to the U.S. Metric Study by Fern E. Hunt and M. Eloise Green of the Ohio State University school of home economics:

| Utensil | Customary Size (inches) | Metric Size (cm) |
|---|---|---|
| Cake Pans | | |
| oblong | 10 x 6 x 1½ | 25 x 15 x 4 |
| | 11 x 7 x 1½ | 28 x 18 x 4 |
| | 12 x 7½ x 2 | 30 x 8 x 5 |
| | 13 x 9 x 2 | 33 x 23 x 5 |
| round | 8 x 1½ | 20 x 4 |
| | 9 x 1½ | 23 x 4 |
| | 10 x 1½ | 25 x 4 |
| square | 8 x 8 x 2 | 21 x 21 x 5 |
| | 9 x 9 x 2 | 23 x 23 x 5 |
| | 10 x 10 x 2 | 25 x 25 x 4 |
| tube | 9 x 3½ | 23 x 9 |
| | 10 x 4 | 25 x 10 |
| Pie Pans & Plates | 4¼ x 1¼ | 11 x 3 |
| | 5 x 1 | 13 x 3 |
| | 6 x 1 | 15 x 3 |
| | 7½ x 1¼ | 19 x 3 |
| | 9½ x 1¼ | 24 x 3 |
| | 10 x 1½ | 27 x 4 |
| | 11 x 1½ | 28 x 4 |
| | 12 x 1½ | 30 x 4 |
| Cookie sheets | 10 x 8 | 25 x 21 |
| | 14 x 10 | 36 x 25 |
| | 15½ x 12 | 39 x 30 |
| | 16 x 11 | 41 x 28 |
| | 17 x 14 | 43 x 36 |
| | 18 x 12 | 46 x 30 |

| Utensil | Customary Size (inches) | Metric Size (cm) |
|---|---|---|
| Jelly Roll Pan | 15½ x 10½ x 1 | 39 x 27 x 3 |
| Loaf Pans | 7½ x 3¾ x 2¼ | 19 x 10 x 6 |
| | 8½ x 4½ x 2½ | 22 x 11 x 6 |
| | 9½ x 5 x 3 | 24 x 13 x 8 |
| | 11 x 7 x 3 | 28 x 18 x 8 |
| | 16 x 4 x 4 | 41 x 10 x 10 |
| Cupcake | 1¾ x 1 | 5 x 2 |
| | 2½ x 1¼ | 6 x 3 |
| | 3 x 1½ | 8 x 4 |

Sauce pans, Dutch ovens and the like will continue to be designated by capacity but the capacity will be stated in liters instead of quarts. Thus, instead of the pint, quart, two quart, five quart and 10 quart pots, the cook will use the half-liter, one liter, two liter, five liter and 10 liter pots.

**5. Cooking Temperatures:** Temperatures will be stated in degrees Celsius instead of Fahrenheit. Exact conversions can be found in the conversion tables elsewhere in this book. A rough rule-of-thumb is that Celsius temperatures are about half those of Fahrenheit. Should the cook have an oven dial or candy or frying thermometer graduated in degrees Celsius, the following table might be useful:

| | Range in degrees Celsius | Range in degrees Fahrenheit |
|---|---|---|
| Oven baking: | | |
| Very slow | 120-135 | 250-275 |
| Slow | 150-165 | 300-325 |
| Moderate | 175-190 | 350-375 |
| Hot | 205-220 | 400-425 |
| Very Hot | 230-245 | 450-475 |
| Roast Beef (with meat thermometer) | | |
| Rare | 60 | 140 |
| Medium | 70 | 160 |
| Well-done | 80 | 170 |
| Deep-frying | | |
| Chicken | 175 | 350 |
| Fish | 175-190 | 350-375 |

| | Range in degrees Celsius | Range in degrees Fahrenheit |
|---|---|---|
| Onions | 190-195 | 375-385 |
| Potatoes | 195-200 | 385-395 |
| | | |
| Candy-making | | |
| Thread | 110-112 | 230-234 |
| Soft ball | 112-115 | 234-240 |
| Firm ball | 118-120 | 244-248 |
| Hard ball | 121-130 | 250-266 |
| Soft crack | 132-143 | 270-290 |
| Hard crack | 149-154 | 300-310 |

# METRICS in the WORKSHOP

Generally, the home handyman (or woman) should have little difficulty in "going metric." The most noticeable change will involve asking the lumber dealer or the hardware salesman for items specified in millimeters instead of inches. For instance, instead of buying a ⅜ths-inch drill bit or screw, one will ask for a 10 mm bit or screw. The 10 mm bit is about $^{1}/_{64}$th inch larger in diameter than the ⅜ths-inch bit but, unless one is doing work requiring extreme precision, the difference should be of little consequence.

The biggest problem will involve wrenches, bolts and nuts. The metric wrenches, bolts and nuts are **not** interchangeable with the customary wrenches, bolts and nuts. For example, a standard customary bolt head is, say, ¾ by ⅞ inches which is about 19.05 mm by 22.225 mm. Standard metric wrenches, however, come only in whole millimeter dimensions. Thus, a 17 mm by 19 mm open-end wrench would be too small to fit the bolt head while the next largest size, 20 mm by 22 mm, is too large in one dimension and too small in the other, and the 21 mm by 23 mm wrench is too large all together.

Similarly, a ¼-inch nut cut with 28 threads per inch cannot be screwed onto a 6 mm bolt with 1.00 thread per millimeter, although at a casual glance, the two may appear to be similar in size. Besides being too large, the threads will not match.

Thus, during the transition to the metric system, the handyperson will have to be careful, making sure that metric wrenches and nuts are matched to metric bolts. A fully-equipped workshop should keep two sets of wrenches and taps and dies, one metric and one customary.

**1. Lumber and board** sizes will be changed to full millimeter dimensions from inches. The size change, however, will be so small as to be unimportant except to builders of large structures. The main reason to make as little change as possible in lumber, board and panel sizes is to minimize any impact on standard building plans and practices. The standard spacing of upright wall studs, for example, is set to accomodate the use of, say, 4-foot by 8-foot wall panels. If the size the wall panels were to be changed significantly, architects and carpenters would have to change the spacing of wall studs. To avoid this kind of complication of "going metric", the sizes of wood building products will be changed by only fractions of a millimeter, at least in the U.S.

The U.S. and Canadian lumber industries are taking slightly different approaches to the switch to metric measurements, however. The U.S. producers are merely converting lumber dimensions from inches to millimeters and then, in effect, rounding off to the nearest whole millimeter. A board that is nominally four inches wide, but actually 3½ inches wide, the new dimension will be 89 mm instead of 88.9 mm. The Canadians are going a bit farther and with some exceptions making the new metric dimension the nearest number of millimeters divisible by five. Hence, the four-inch wide board in Canada will be changed to 90 mm.

A major gain in the change to metric dimensions in wood products will be that the consumer no longer will have to remember that the actual or "net" size of a board is smaller than the "nominal" size. As many a home repairman has discovered to his dismay, the ubiquitous 2 x 4 isn't two inches by four inches but is actually 1½ inches by 3½ inches. The new metric dimensions, in millimeters, be the actual size of the board, at least to the nearest whole millimeter. The metric equivalent of the 2 x 4 will be a 38 x 89 (or 38 x 90 in Canada), which is the board's actual or "net" dry dimensions in millimeters.

A complete listing of the new recommended metric dimensions in softwood lumber products has been agreed upon by the American National Metric Council and can be obtained through such industry associations as the Western Wood Products Association in Portland, Ore. The listings cover such lumber products as flooring, ceiling, stepping, shiplap, timbers, siding and tongue-and-grooved lumber.

Listed below are some of the most widely used lumber and board dimensions. Canadian sizes are shown only when different from U.S. sizes.

| Nominal (in) | Net, dry (in) | Metric (mm) U.S. | Canada |
|---|---|---|---|
| | | **Thicknesses** | |
| 3/8 | 5/16 | 8 | |
| 1/2 | 7/16 | 11 | |
| 5/8 | 9/16 | 14 | |
| 3/4 | 5/8 | 16 | |
| 1 | 3/4 | 19 | |
| 1 1/4 | 1 | 25 | |
| 1 1/2 | 1 1/4 | 32 | |
| 1 3/4 | 1 3/8 | 35 | |
| 2 | 1 1/2 | 38 | |
| 2 1/2 | 2 | 51 | 50 |
| 3 | 2 1/2 | 64 | 65 |
| 3 1/2 | 3 | 76 | 75 |
| 4 | 3 1/2 | 89 | 90 |
| | | **Face Widths** | |
| 2 | 1 1/2 | 38 | |
| 3 | 2 1/2 | 64 | 65 |
| 4 | 3 1/2 | 89 | 90 |
| 5 | 4 1/2 | 114 | 115 |
| 6 | 5 1/2 | 139 | 140 |
| 7 | 6 1/2 | 165 | 165 |
| 8 | 7 1/4 | 185 | 190 |
| 9 | 8 1/4 | 210 | 215 |
| 10 | 9 1/4 | 235 | 240 |
| 11 | 10 1/4 | 260 | 265 |
| 12 | 11 1/4 | 285 | 290 |
| 14 | 13 1/4 | 335 | 340 |
| 16 | 15 1/4 | 385 | 390 |

Panels & Modules

| Inches | Metric (mm) |
|---|---|
| 4 x 4 | 100 x 100 |
| 12 x 12 | 300 x 300 |
| 16 x 16 | 400 x 400 |
| 24 x 24 | 600 x 600 |
| 48 x 48 (4′ x 4′) | 1200 x 1200 |
| 48 x 96 (4′ x 8′) | 1200 x 2400 |

**2. Screws and bolts:** If the change to the metric system does nothing else, it will greatly simplify the workshop's—and the nation's—inventory of screws and bolts. Over the past several decades, literally thousands of different sizes and grades of threaded fasteners have come into use. There are more than 200 different sizes of wood screws alone, ranging from two different lengths of the tiny #0 screw to 17 lengths of the #14 screw (not to mention the fact that each size comes with flat, oval or round heads). One almost has to be an experienced carpenter to remember that a #14 screw is 15/64 inch in diameter while the next largest diameter screw, the #16, is 9/32 inch in diameter. If one also has to deal with machine bolts, stove bolts, sheet metal screws, hex bolts, cap screws, etc., the size situation becomes almost hopeless.

Unlike other areas of commerce, it isn't possible to restate the dimensions of threaded fasteners in metric units, i.e. millimeters, instead of inches. Diameters and lengths would come out impractical fractions of millimeters and the pitch of threads—the distance between crests—would be almost impossible to restate in millimeters. For example, 5/16 inch diameter screw with 24 threads per inch would come out to be 7.937 500 mm diameter with a pitch of 1.058 333 333 333 mm.

Rather than try to make a metric substitute for each of the thousands of different sized bolts and screws currently being used, groups such as the Industrial Fastener Institute, have agreed to use the metric changeover to reform the entire system of fastener sizes. As a result of about 90% of the sizes of fasteners will be eliminated. In other words, for every 100 different sizes currently used there will be only 10 standard metric sizes. The reductions in inventories alone should more than pay for the cost of switching fasteners to the metric system.

Under the metric system, U.S. screws and bolts will have standard diameters stated in millimeters preceded by the letter M. Standard lengths also will be stated in millimeters divisable by five for lengths up to 100 mm, divisable by 10 for lengths of 100 mm to 200 mm, and divisable by 20 for lengths beyond 200 mm.

Thus, the handyman accustomed to asking the hardware clerk for a #5 screw (1/8 inch diameter) by one inch long will now ask for an M-3 by 25 meaning, of course, a 33 mm diameter screw 25 mm long (which, incidentally, is slightly smaller than #5 screw but will be the nearest metric equivalent).

The standard metric diameters for fasteners in relation to customary inch diameters are as follows:

| Metric (mm) | Nearest (in) |
|---|---|
| M-1.6 | 1/16 |
| M2 | 5/64 |
| M-2.5 | 3/32-7/64 |
| M-3 | 1/8 |
| M-3.5 | 9/64 |
| M-4 | 5/32 |
| M-5 | 3/16-13/64 |
| M-6 | 15/64 |
| M-8 | 5/16 |
| M-10 | 3/8-13/32 |
| M-12 | 15/32-1/2 |
| M-14 | 35/64-37/64 |
| M-16 | 5/8-21/32 |
| M-20 | 3/4-7/8 |
| M-24 | 15/16-1 inch |
| M-30 | 13/16-1¼ |
| M-36 | 1½ |

Standard metric lengths of fasteners beginning at 5 mm (About 1/5 inch) come in increments of 5 mm up to 100 mm (that is, 5, 10, 15, 20, 25, etc.). Fasteners of 100 mm to 200 mm (approximately 4 to 8 inches) come in standard increments of 10 mm (i.e. 110, 120, 130, etc.) while those longer than 200 mm come in standard increments of 20 mm (i.e. 220, 240, 260, etc.).

**3. Drill Bits:** Metric twist drill bits are becoming available in the U.S. although not in the variety of sizes available in customary dimensions. Customary drill bits are made in increments of 1/64th inch beginning as small as 1/16th inch diameter. Metric drill bits are made in increments of tenths of a millimeter. Since the woodworker usually wants to drill holes that are fractionally smaller than the diameter of the woodscrew or fractionally larger than the diameter of a bolt, the metric drill bits that are most used and thus most easily available are a few tenths of a millimeter smaller or larger than standard metric screws and bolts. In picking the right sized metric bit, it might help to remember that 0.4 mm is about 1/64th inch.

An example of the metric twist bits available for electric drills is drawn from a recent Sears, Roebuck & Co. catalog.

| Metric Bit (mm) | Nearest Customary Bit (in) |
|---|---|
| 3.3 | 1/8-9/64 |
| 4.1 | 5/32-11/64 |
| 4.2 | 5/32-11/64 |
| 5.0 | 13/64 |
| 6.0 | 15/64 |
| 6.7 | 17/64 |
| 7.7 | 19/64-5/16 |
| 8.0 | 5/16-21/64 |
| 8.5 | 21/64-11/32 |
| 8.7 | 11/32 |
| 9.0 | 23/64 |
| 10.5 | 13/32-27/64 |
| 12.0 | 15/32-31/64 |
| 12.8 | 1/2 |
| 14.0 | 35/64 |

**4. Taps and Dies:** As noted above, metric and customary screw threads simply cannot be made to match. Customary screw threads are machined on the basis of a standard number of threads per inch while metric screws are machined on the basis of a standard number of threads (or fractions of a thread) per millimeter and never the twain shall meet. Thus, a well-equipped workshop should have a set of metric taps and dies.

Below is a list of metric thread sizes of taps and dies offered recently by Sears, Roebuck & Co. The diameters range from 3 mm (slightly smaller than ⅛ inch) to 18 mm (slightly smaller than ¾ inch). The other dimension is the thread pitch—the distance between crests—stated in millimeters. This differs from customary taps and dies where thread sizes are stated in threads per inch. The nearest customary thread sizes are given only for guidance since customary and metric threads are not interchangeable. Customary thread sizes are stated by number in the case of machine screw threads and in fractions of an inch for National Coarse and Fine Threads.

Note: For pipes, metric tap and die sets use British Standard Pipe threads whereas customary pipe taps and dies are set for National Pipe Threads.

| Metric | | | Nearest Customary | |
|---|---|---|---|---|
| Dia. (mm) | Pitch (mm) | Approximate Threads per In. | Dia. (# or in.) | Threads Per In. |
| 3 | 0.50 | 50.8 | # 4 | 48 |
| 3 | 0.60 | 42.3 | # 4 | 40 |
| 4 | 0.70 | 36.3 | # 8 | 36 |
| 4 | 0.75 | 33.9 | # 8 | 32 |
| 5 | 0.80 | 31.75 | #10 | 32 |
| 5 | 0.90 | 28.2 | #10 | 24 |
| 6 | 1.00 | 25.4 | 1/4 | 28 |
| 7 | 1.00 | 25.4 | — | — |
| 8 | 1.25 | 20.3 | 5/16 | 18 |
| 9 | 1.00 | 25.4 | 3/8 | 24 |
| 9 | 1.25 | 20.3 | — | — |
| 10 | 1.25 | 20.3 | — | — |
| 10 | 1.50 | 16.9 | 3/8 | 16 |
| 11 | 1.50 | 16.9 | 7/16 | 20 |
| 12 | 1.50 | 16.9 | 1/2 | 20 |
| 12 | 1.75 | 14.5 | 1/2 | 13 |
| 14 | 1.25 | 20.3 | 9/16 | 18 |
| 14 | 2.00 | 12.7 | 9/16 | 12 |
| 16 | 1.50 | 16.9 | 5/8 | 18 |
| 16 | 2.00 | 12.7 | 5/8 | 11 |
| 18 | 1.50 | 16.9 | 3/4 | 16 |
| 18 | 2.50 | 10.2 | 3/4 | 10 |

**5. Wrenches:** Except for the obvious case of adjustable wrenches, customary-sized wrenches can't be used on metric-sized bolt heads and metric-sized wrenches can't be used on customary-sized bolts. Hence, during the transition period two complete sets of open-end, box, socket and hex-key wrenches may be needed.

Below are examples of standard metric-sized wrenches in relation to the nearest sized customary wrenches.

Hex key wrenches come in standard metric sizes of 2, 2.5, 3, 4, 5, 6, 7 and 8 mm and, for drain plugs, 12 mm and 17 mm.

Socket wrenches come in metric sizes beginning at 4 mm (about $^{5}/_{32}$ inch) and increase in 1 mm increments. Customary socket wrenches begin at $^{5}/_{32}$ inch and range upward in $^{1}/_{32}$ inch increments. Fortunately, handle sizes for metric sockets are the same as for customary sockets.

| Metric (mm) | Nearest Customary (in) |
|---|---|
| 6 x 8 | 1/4 x 5/16 |
| 7 x 9 | |
| 10 x 11 | 3/8 x 7/16 |
| 12 x 14 | 1/2 x 9/16 |
| 13 x 15 | 9/16 x 5/8 |
| 16 x 18 | 5/8 x 3/4 |
| 17 x 19 | 11/16 x 13/16 |
| 20 x 22 | 3/4 x 7/8 |
| 21 x 23 | |
| 22 x 24 | |
| 24 x 26 | 15/16 x 1 |
| 25 x 27 | |
| 28 x 30 | 1 1/16 x 1 1/8 |
| 30 x 32 | |

# METRICS AT THE OFFICE

The only noticeable change resulting from metrication in the office will be in specifying sizes of stationery, envelopes, file cards and other paper products in millimeters instead of inches. The actual size of these paper products will remain the same, at least during the first several years of transition to the metric system. As a result, present-day office machines—typewriters, copying, adding, dictation, teleprinters, etc.—will not have to be modified to accommodate metric-dimension paper.

The American Paper Institute has recommended that labels on **printing and writing papers** contain both customary and metric measurements, that is, length and width should be stated in millimeters rounded off the nearest whole millimeter as well as inches and "basis weights" should be stated in grams per square meter as well as pounds per ream of 500 sheets.

Examples:
- 8½″ x 11″ will be 216 mm x 279 mm
- 8½″ x 14″ will be 216 mm x 356 mm
- 3″ x 5″ will be 76 mm x 127 mm
- 4*″ x 9½″ will be 100 mm x 240 mm

**Basis weights** of paper will be converted to **grammage** or grams per square meter. To do this, take the basis weight, stated in pounds per ream, multiply by 1 406.5 and divide by the number of square inches in the sheet. For example, 17″ x 22″ paper that weighs 20 lb. per ream will have a grammage of 75 or 75 $g/m^2$ (20 lb. times 1 406.5 divided by 374 square inches).

An example of the grammage of paper that comes from the mill in 17″ x 22″ sheets (equal to four letter-sized sheets) is as follows:

| Basis Weight (lb./ream) | Grammage ($g/m^2$) |
|---|---|
| 9 | 34 |
| 11 | 41 |
| 13 | 49 |
| 15 | 56 |
| 16 | 60 |
| 20 | 75 |
| 24 | 90 |
| 28 | 105 |
| 32 | 120 |
| 36 | 135 |
| 40 | 150 |

Note: The grammage of other standard sizes of paper can be determined by the formula stated above or by multiplying the basis weight by the following conversion factors:

| | | |
|---|---|---|
| 17″ x 22″ | multiply basis weight by | 3.7596 |
| 20″ x 26″ | ″ | 2.7042 |
| 20″ x 30″ | ″ | 2.3442 |
| 24″ x 36″ | ″ | 1.6275 |
| 25″ x 38″ | ″ | 1.4802 |

# Definitions and Conversion Factors

## A

**ab-** Prefix used when stating electromagnetic measurements in the centimeter-gram-second system instead of the SI system.
**abampere** Unit of electric current equal to 10 amperes.
**abcoulomb** Unit of electric charge equal to 10 coulombs.
**abfarad** Unit of capacitance equal to 1 000 000 000 farads.
**abhenry** Unit of inductance equal to 0.000 000 001 henry.
**abohm** Unit of electrical resistance equal to 0.000 000 001 ohm.
**abvolt** Unit of electrical force equal to 0.000 000 010 volt.
**acre** Unit of area equal to 4 046.856 square meters or 43 560 square feet.
**acre-foot** Volume of a liquid, usually water, covering one acre to a depth of one foot, equal to 1 233.482 cubic meters.
**ampere** Unit of electric current which, if maintained in two parallel conductors one meter apart, produces a force equal to 0.000 000 200 newton per meter of length.
**ampere-turn** Unit of magnetic force produced by a current of one ampere in one turn of a closed loop.
**angstrom** Unit of length, usually used in measuring light waves, equal to 0.000 000 000 1 m (0.1 nanometer) or 0.000 000 003 937 007 inch (approximately four-billionths of an inch).
**are** Metric unit of area equal to 100 square meters or 1 076.390 square feet; 100 are equal one hectare.
**astronomical unit** A unit of celestial distances equal to the mean distance between the sun and the earth or 149 598 000 kilometers or 92 955 870 statute miles.

**atmosphere, physical** Unit of pressure equal to the air pressure at mean sea level or 1.033 227 kilograms-force per square centimeter or 101.325 kilopascals or 14.695 940 pounds-force per square inch.

**atmosphere, technical** A pressure of one kilogram-force per square centimeter, or 98.066 kilopascals.

**avoirdupois** A system of weights based on 16 drams to the ounce and 16 ounces to the pound, from Old French meaning "goods sold by weight."

## B

**bar** Unit of pressure equal to 100 kilopascals or 100 000 newtons per square meter or 14.5 pounds-force per square inch.

**barrel** Unit of liquid volume usually equal to 31.5 gallons (119.24 liters) except in the petroleum industry where one barrel equals 42 gallons (158.987 liters); may also be different in other commodities.

**board foot** Unit of volume used in measuring lumber, equal to a board one foot square and one inch thick or 144 cubic inches or 0.002 359 737 cubic meters (2 359.737 cubic centimeters).

**Btu (British thermal unit)** A unit of heat energy required to raise one pound of water one degree Fahrenheit at sea level (or 1/180 of the heat required to raise one pound of water from 0 degrees C to 100 degrees C at sea level); equal to 1 055 joules or 0.252 kilocalories.

**BTU per hour** A unit of power equal to 0.293 watt or approximately 0.000 4 horsepower.

**BTU per second** A unit of power equal to 1 054.8 watts (1.054 8 kilowatts).

## C

**cable length** A marine unit of distance variously defined as equal to 600 feet (183 meters) or equal to 720 feet or 120 fathoms (219.5 meters).

**calorie** amount of heat energy required to raise the temperature of one gram of water one degree C; a technical calorie is equal to 4.186 501 joules while a thermochemical calorie is equal to 4.184 joules. Often expressed in kilocalories (one kilocalorie is approximately four BTUs).

**calories per hour** A unit of power equal to 0.001 163 watt.

**candela** The basic unit of light intensity defined as the luminous intensity 1/60 of a square centimeter of the surface of a black body at the solidification temperature of platinum (2 042 degrees K).

**carat** Unit of weight used in measuring gems; equal to 200 milligrams.

**centimeter** Unit of length equal to 0.01 meter or 0.393 700 inches.

**centimeter of mercury** A unit of pressure equal to the pressure required to support a column of mercury one centimeter high; equal to approximately 1.333 kilopascals or 135.996 kilograms-force per square meter or 0.193 333 pound per square inch or 27.854 pounds per square foot. One centimeter of mercury equals 0.393 700 inch of mercury.

**centimeter of water** The pressure required to support a column of water one centimeter high, equal to one kilogram-force per square meter or approximately 9.8 pascals or 1 421.3 pounds per square inch or approximately 0.2 pounds per square foot.

**centimeter per second** A unit of velocity equal to 0.032 808 feet per second or 0.393 700 inch per second or 0.036 kilometer per hour.

**centimeter per second squared** A unit of acceleration equal to 0.032 808 foot per second per second or 0.393 700 inch per second per second.

**centipoise** A unit of absolute viscosity equal to 0.01 poise or 0.001 newton-second per square meter, or 0.208 854 pound-second per square foot.

**centistokes** A unit of kinematic viscosity equal to 0.01 stokes or 387.5 square feet per hour or 0.107 639 square feet per second.

**chain** A surveyor's measure of length equal to 100 links of 7.92 inches each or a total length of 66 feet, equal to 20.1168 meters.

**cord** A stack of wood four feet by four feet by eight feet or 128 cubic feet, equal to 3.624 556 cubic meters.

**cord foot** A stack of wood equal to 16 cubic feet or about 0.453 070 cubic meters.

**coulomb** The amount of electricity transported by a current of one ampere in one second.

**cubic centimeter** Unit of dry volume equal to 0.061 024 cubic inch, or a unit of liquid volume equal to 0.001 liter or 0.002 113 pint or 0.338 fluid ounce.

**cubic decimeter** Unit of dry volume equal to 0.035 315 cubic feet or 61.02 cubic inches; a unit of liquid volume equal to one liter or 0.264 172 gallon or 1.057 quarts.

**cubic foot** Unit of dry volume equal to 0.028 317 cubic meter.
**cubic inch** Unit of dry volume equal to 16.387 cubic centimeters.
**cubic meter** Unit of volume equal to 35.314 660 cubic feet or 1.307 950 cubic yards; in liquid volume equal to 1 000 liters.
**cubic yard** Unit of volume equal to 27 cubic feet or 0.764 555 cubic meter.
**cubit** Unit of length equal to 18 inches; a biblical cubit is 21.8 inches.
**cup** Unit of volume usually used in cooking equal to 16 tablespoons, or 236.6 milliliters (0.236 6 liters). A metric cup is slightly larger and is equal to 250 milliliters. Whereas four U.S. cups equal one quart, four metric cups equal one liter.

## D

**day** Length of time equal to one rotation of the earth but since the earth's rotation is independent of the earth's movement around the sun, more precise definitions have been developed. The day is 86,400 seconds and the second is defined in terms of the vibrations of the cesium atom. The mean solar day is 1/365.242 198 79 part of the mean solar year which, in turn, is based on the length of the tropical year 1900. The sidereal day is based on the earth's rotation in relation to the stars rather than the sun. In sidereal time, the mean solar day is 24 hours, 03 minutes, 56.5 seconds.
**decimeter** Unit of length equal to 0.1 meter or 3.937 inches.
**degree, angular** Unit for measurement of angles in a plane, equal to 1/360th of a circle or 0.017 453 radian, or 60 minutes or 3 600 seconds of arc.
**degree Celsius** Unit used for measuring temperature, formerly called degree centigrade, equal to 1/100th of the difference between the freezing and boiling points of water. Specifically defined as being equal to one kelvin, the basic temperature unit of the metric system. To convert kelvin to degrees Celsius add 273.15. To convert Fahrenheit to Celsius subtract 32 and multiply the result by 5/9 or 0.555.
**degree, centigrade** see degree, Celsius.
**degree Fahrenheit** Unit of temperature equal to 1/180 of the difference between the freezing and boiling points of water when the freezing point is set at 32 degrees and the boiling point at 212 degrees (zero degrees is the freezing point of a certain mixture of ice, water and salt). To convert Celsius temperatures to Fahrenheit multiply by 9/5 or 1.8 and add 32; to convert kelvin

temperatures to Fahrenheit subtract 255.37 and multiply by 9/5 or 1.8.

**dram** A unit of weight. In the apothecaries' or troy systems, equal to 60 grains or 3.887 935 grams; in the avoirdupois system equal to 1/16 of an ounce or 27 11/32 grains or 1.771 845 grams.

**dyne (obsolete)** A unit of force which gives one gram of mass an acceleration of one centimeter per second square; equal to 0.000 010 newton or about 0.000 002 pound-force.

**dyne per cubic centimeter** A unit weight equal to 0.063 659 pound-force per cubic foot.

**dyne-centimeter** An erg.

## E

**erg** Unit of energy or work done by a force of one dyne displacing a point one centimeter; equal to 0.000 000 1 joule or 0.000 000 074 foot-pound.

## F

**farad** The basic unit of capacitance equal to the capacitance of a capacitor with a difference of potential between its plates of one volt when charged by one coulomb of electricity. One farad equals one ampere-second (one coulomb) divided by one volt.

**fathom** Marine unit of length equal to six feet or 1.828 800 meters.

**fluid ounce** A unit of liquid volume equal in the U.S. system to 29.573 milliliters and in the British system to 28.416 milliliters.

**fluidram** A unit of volume equal in the U.S. system to 3.696 milliliters and in the British system to 3.551 6 milliliters.

**foot** A unit of length in the FPS system equal to 0.3048 meter or 30.48 centimeters.

**foot candle** A unit of illumination equal to the illumination of all points that are one foot distant from a light source of one candela.

**foot per hour** A unit of velocity equal to 0.008 466 centimeters per second.

**foot per minute** A unit of velocity equal to 0.016 666 foot per second or 0.005 080 meter per second.

**foot per minute squared** A unit of acceleration equal to 0.000 278 foot per second per second or 0.000 084 666 meter per second per second or 0.304 800 meter per minute per minute.

**foot per second** A unit of velocity equal to 0.304 800 meter per second or 0.681 818 mile per hour or 1.097 280 kilometers per hour.

**foot per second squared** A unit of acceleration equal to 0.304 800 meters per second per second.

**foot-pound** A unit of energy or work done by a pound of force moving through one foot; equal to 0.001 285 Btu or 1.355 818 joules or 0.138 255 meter-kilogram.

**foot-poundal** Unit of energy or work done by a force of one poundal moving through one foot; equal 0.042 140 joule.

**foot-pound per hour** A unit of power equal to 0.000 376 616 watt.

**foot-pound per minute** A unit of power equal to 0.001 285 347 Btu or 0.022 597 watt.

**furlong** Unit of length equal to 40 rods or 1/8 mile (660 feet) or 201.168 meters.

## G

**gallon** Unit of volume. In the U.S., one gallon liquid measure is equal to 231 cubic inches or 3.785 412 liters and one gallon dry measure is equal to 0.004 405 cubic meter (or 0.155 557 cubic foot). In the British system, a gallon is the volume occupied by 10 pounds of water and is equal to about 1.2 U.S. liquid gallons; also equal to 0.004 546 cubic meter or 277.420 cubic inches.

**gamma** A unit of magnetic field strength equal to 0.000 01 gauss or 0.000 000 001 telsa where one telsa equals one weber (or one volt-second) per square meter.

**gauss** A unit of magnetism equal to 0.000 100 telsa or 0.000 100 weber (or volt-second) per square meter. An abtelsa.

**gilbert** Unit of magnetic force equal to 0.795 775 ampere turn.

**gill** Unit of volume equal to ¼ pint liquid. In the U.S., a gill equals 118.294 milliliters. In Britain, a gill equals 142.066 milliliters.

**grad** Angular measurement equal to 0.9 degree or 54 minutes.

**grain** A unit of weight originally based on the weight of a grain of wheat; equal to 0.064 799 gram (64.8 milligrams) or 0.002 083 troy (apothecaries') ounce or 0.002 285 avoirdupois ounce.

**gram** A metric unit of weight equal to 0.001 kilogram or 0.035 274 avoirdupois ounce or 0.032 151 troy ounce.

**gram per cubic centimeter** A unit of density equal to 0.036 127 pound per cubic inch.

**gram-force** A unit of force equal to 0.002 204 622 pound-force or 0.009 806 650 newton.

### H

**hectare** A metric unit of area equal to 100 ares or 10,000 square meters; equal to 107 639.1 square feet or 2.471 054 acres.

**henry** An electrostatic unit of inductance when an electric current in a circuit changes at the rate of one ampere per second and induces a force of one volt. One henry equals one joule per ampere second.

**hogshead** A unit of volume equal to two barrels or 63 gallons or 238.48 liters or 0.238 480 cubic meter.

**horsepower** A unit of power. In the customary foot-pound-second (FPS) system one horsepower equals 550 foot-pounds per second or 745.699 watts. A "metric" horsepower (a non-SI unit) equals 542.448 foot-pounds per second or 735.498 800 watts.

**horsepower, boiler** Unit of power equal to 13.337 210 metric horsepower or 9 809.5 watts.

**horsepower, electric** Unit of power equal to 1.014 278 metric horsepower or 746 watts.

**horsepower, water** Unit of power equal to 1.014 336 metric horsepower or 746.043 watts.

**hundredweight, long** Unit of weight equal to 112 pounds (0.05 long ton) or 50.802 350 kilograms.

**hundredweight, short** Unit of weight equal to 100 pounds or 45.359 230 kilograms.

### I

**inch** Unit of length equal to 2.54 centimeters.

**inch of mercury** Unit of pressure equal to 3.386 389 kilopascals or 345.315 500 kilogram-force per square meter or 0.491 155 pounds per square inch.

**inch of water** Unit of pressure equal to 0.249 082 kilopascals or 25.399 290 kilograms-force per square meter or 0.036 126 pound per square inch.

**inch per second** Unit of velocity equal to 0.025 400 meter per second.

**inch per second squared** Unit of acceleration equal to 0.254 400 meter per second per second or 0.083 333 foot per second per second.

## J

**joule** Basic unit of energy or work in the metric system, equal to the work done by a force of one newton moving through one meter; equal to 0.101 972 meter-kilogram-force or 0.737 562 foot-pound-force or 0.000 947 867 Btu.

## K

**kelvin** The basic unit of temperature measurement used in the metric system, defined as 1/273.16 of the temperature at which water can exist simultaneously as vapor, liquid and solid (the triple point of water). Zero kelvin or absolute zero is the point at which a body would have no heat energy and all molecular motion is stopped. To convert Fahrenheit temperatures to kelvin add 459.67 and multiply by 5/9 or 0.555. To convert Celsius to kelvin add 273.15.

**kilogram** Basic unit of mass in the metric system equal to the mass of a cylinder of platenium-iridium stored in a vault at Sevres, France. One kilogram equals 2.204 622 pounds.

**kilogram per cubic centimeter** A unit of density equal to 62 428 pounds per cubic foot or 36.127 200 pounds per cubic inch.

**kilogram per cubic meter** Unit of density equal to 0.062 428 pound per cubic foot or 0.000 036 127 pound per cubic inch.

**kilogram per meter-second** Unit of viscosity equal to 0.671 969 pound per foot second or 10 poise.

**kilogram per square centimeter** Unit of pressure equal to one technical atmosphere or 0.967 841 5 physical atmosphere or 14.695 94 pounds per square inch or 98 066.5 newtons per square meter (98.066 5 kilopascals).

**kilogram-force** Unit of force equal to 9.806 650 newtons or 2.204 622 pounds-force.

**kilogram-force-meter** Unit of moment equal to 7.233 016 pound-force-foot

**kilometer** Unit of length equal to 1000 meters or 0.621 371 mile or 3 280.839 feet.

**kilometer per hour** Unit of velocity equal to 0.621 371 mile per hour or 0.911 344 foot per second or 0.277 777 meter per second or 0.540 knot.

**kilopound** Unit of weight equal to 1 000 pounds.

**kilopound per cubic foot** Unit of density equal to 16 018.460 kilograms per cubic meter.

**kilopound-force** Unit of force equal to 4 448.222 newtons or 453.592 300 kilogram-force.

**kilowatt** Unit of power equal to 1 000 watts.

**kilowatt-hour** Unit of energy or work equal to 3 600 000 joules (3.6 megajoules) or 2 655 222 foot-pounds or 860.420 kilocalories.

**knot** Unit of velocity equal to one nautical mile per hour or 1.150 696 statute miles per hour or 1.852 kilometers per hour.

## L

**light year** Unit of astronomical distance; the distance traveled in one year by light traveling at a velocity of 299 792.4580 kilometers per second; equal to a distance of about 9 460 550 000 000 kilometers or 5 878 512 000 000 miles.

**link** Unit of length in surveyors' measure equal to 7.92 inches or 20.116 800 centimeters; 25 links equal one rod.

**liter** Metric unit of volume equal to 0.001 cubic meter (one cubic decimeter) or 0.264 172 U.S. fluid gallon or 0.227 021 U.S. dry gallon, or 1.0567 liquid quarts or 0.908 dry quart.

**lumen** A unit for measuring the rate of flow of the visible part of the energy emitted by a light source, specifically, the luminous flux or flow of visible light within an angle of one steradian from a light source having an intensity of one candela. Also defined as 1/680th watt of yellow-green light of a wavelength of 550 nanometers.

## M

**MKS** or **MKSA** Abbreviation for "meter-kilogram-second-ampere" system as opposed to the FPS or "foot-pound-second" system of measurement. MKS was, for many years, the "metric system", having replaced the older "centigrade-gram-second" system after World War II. The MKS system is being replaced worldwide by the International System of Units or SI (after the French name, *Systeme International d'Unites)* which uses such basic units as the Kelvin temperature scale and the candela.

**maxwell** A unit of magnetic flux in the old centimeter-gram-second system, now known as abweber; equal to 0.000 000 01 weber. Defined as the magnetic flux which produces a force of one abvolt in a circuit of one turn as the flux is reduced to zero in one second.

**meter** The basic metric unit of length. Originally defined as the length of a metal bar stored in a vault at Sevres, France, but now defined as the length equal to 1 650 763.73 wavelengths of orange-red light emitted by the krypton-86 atom as it changes between certain energy levels. A meter is equal to 39.370 070 inches or 3.280 840 feet.

**meter-kilogram** A unit of energy or work done by one kilogram of force displacing a point one meter; equal to 9.806 652 joules or 1.355 819 foot-pounds.

**meter per second** A unit of velocity equal to 3.280 839 feet per second or about 3.6 kilometers per hour.

**mile** A unit of length originally equal to 1 000 paces. A statute or land mile is equal to 5 280 feet or 1.609 344 kilometers. A nautical mile is equal to one minute of longitude at the equator (or any Great Circle of the earth) or 1 852 meters (1.852 kilometers); equal to 1.150 780 statute miles.

**mile (naut) per hour** Unit of velocity equal to 1.852 kilometers per hour or one knot.

**mile (stat) per hour** Unit of velocity equal to 1.609 344 kilometers per hour or 0.447 040 meter per second or 1.466 667 feet per second or 0.869 039 knot.

**mile (stat) per minute** Unit of velocity equal to 26.822 240 meters per second.

**mile (stat) per second** Unit of velocity equal to 1 609.344 meters per second.

**millimeter** Unit of length equal to 0.001 meter or 0.039 370 inch.

**minim** Unit of liquid volume equal to 1/60 of a dram, originally one drop. In U.S. a minim equals 0.061 610 milliliter; in Britain a minim equals 0.059 194 milliliter.

**minute, angular** Equal to 1/60th or 0.016 666 of a degree; or 0.000 290 888 radian or 0.018 518 510 grad.

**minute, time** 1/60th or 0.016 666 of an hour; one day equals 1 440 minutes.

**mole** A unit of an amount of a substance, usually used in chemistry; defined as the amount of substance composed of the same number of particles—atoms, molecules, ions or other particles—as there are atoms in 0.012 kilogram of pure carbon-12.

## N

**nail** A British unit of length used in measuring cloth, equal to 2.25 inches; 16 nails equals one yard; equal to 5.715 centimeters.

**newton** The unit of force, equal to the force needed to accelerate one kilogram at a rate of one meter per second squared; equal to 100 000 dynes or 7.233 011 poundals or 0.224 809 pound-force or 0.101 972 kilogram-force.

**newton per cubic meter** A unit weight equal to 0.006 365 883 pound per cubic foot or 0.101 972 kilogram per cubic meter.

**newton-meter** A unit of moment of force equal to 8.850 746 pound-inches or 0.737 562 pound-foot or 0.101 972 kilogram-meter.

## O

**ohm** The unit of electrical resistance equal to the resistance when a potential of one volt between two points on a conductor produces a current of one ampere; one ohm equals one volt per ampere or one joule-second per coulomb-squared or one watt per ampere-squared.

**ounce** A unit of weight originally 1/12th of a Roman pound. An avoirdupois ounce is equal to 16 drams or 437.5 grains or 28.349 520 grams. A troy or apothecaries' ounce is equal to 8 drams or 480 grains or 31.103 480 grams. It requires 16 avoirdupois ounces to equal one avoirdupois pound (0.453 592 kilograms) while it requires 12 troy ounces to equal one troy pound (0.373 242 kilograms). In liquid measure a U.S. fluid ounce equals 29.573 milliliters.

**ounce-force** A unit of force equal to 0.278 014 newton or 0.028 349 520 kilogram-force.

## P

**parsec** A unit of astronomical distance equal to 3.258 light years or about 19.16 trillion miles or 30.83 trillion kilometers.

**pascal** The unit of pressure or stress equal to a pressure of one newton per square meter. In the SI pascals will be used instead of bars in pressure measurements; one bar equals 100 kilopascals. One pascal equals 0.000 145 038 pound per square inch or 0.000 010 197 kilogram per square centimeter.

**peck** A unit of capacity equal to eight quarts or ¼ bushel. In the U.S. a peck equals 0.008 809 768 cubic meter while in Britain a peck equals 0.009 092 180 cubic meter.

**pennyweight** A unit of troy weight equal to 24 grains or 0.05 troy ounce or 1.555 grams.

**pint** A unit of capacity equal to four gills or ½ quart. In U.S. a liquid pint is equal to 28.875 cubic inches or 473 milliliters while a dry pint is equal to 33.6 cubic inches or 550.610 cubic centimeters. In Britain a pint equals 34.678 cubic inches or 568.261 cubic centimeters.

**poise** A unit of absolute viscosity in the SI equal to one dyne-second per square centimeter or 0.1 newton-second per square meter. Also equal to 0.067 197 pound per foot-second or 0.1 kilogram per meter-second, or 0.010 197 kilogram-second per square meter or 0.002 089 pound-second per square foot.

**pound** A unit of weight. In avoirdupois weight, one pound equals 16 ounces or 7 000 grains or 0.453 592 kilogram. In troy or apothecaries' weight one pound equals 12 ounces or 5 760 grains or 0.373 242 kilogram.

**pound per cubic foot** A unit of density equal to 16.018 460 kilograms per cubic meter or 0.000 578 703 pound per cubic inch.

**pound per cubic inch** A unit of density equal to 27.679 910 grams per cubic centimeter or 1 728 pounds per cubic foot.

**pound per foot-second** A unit of viscosity equal to 1.488 164 kilogram per meter-second or 14.881 640 poise or 0.151 750 kilogram-second per square meter or 0.031 081 pound-second per square foot. Also equal to 0.000 215 840 reyn.

**pound per square foot** Unit of pressure equal to 47.880 260 pascals or 4.882 428 kilograms per square meter.

**pound per square inch** Unit of pressure equal to 6 894.757 pascals (newtons per square meter) or 0.068 947 57 bar or 0.070 301 kilogram per square centimeter.

**poundal-foot** See foot-poundal.

**pound-foot** see foot-pound.

**pound-force** A unit of force equal to 4.448 222 newtons or 0.453 592 kilogram-force. A troy pound-force is equal to 3.660 251 newtons or 0.822 569 avoirdupois pound-force.

## Q

**quart** Unit of capacity or volume equal to ¼ of a gallon. In the U.S. a liquid quart is equal to 0.946 353 liter and a dry quart is equal to 1 101 cubic centimers (1.101 liters). In Britain, a quart in both liquid and dry measure is equal to 1.136 520 liters.

**quarter** Unit of weight or mass equal to ¼ short ton or 500 pounds or 226.796 185 kilograms. In troy weight, a quarter is ¼ hundredweight or 25 troy pounds or 9.331 043 kilograms. In Britain, a quarter is a unit of volume equal to eight bushels or 0.290 950 cubic meter.

## R

**radian** A plane angle, formed by two radii of a circle, with an arc length equal to the radius. One radian is equal to 57.295 78 degrees or 3 437.747 minutes or 206 264.8 seconds or 63.662 grads. Also a radian is equal in degrees to 180 divided by *pi* or, in grads, to 200 divided by *pi*.

**reyn** A unit of absolute viscosity in the FPS system equal to 144 pound-seconds per square foot. Also equal to 6 894.756 kilograms per meter-second or 703.069 5 kilogram-seconds per square meter or 14.881 640 poise.

**rod** A surveyor's measure of length equal to 25 links; equal to 5.5 yards or 16.5 feet or 5.029 200 meters.

## S

**SI** Abbreviation for *System International d'Unites* or International System of Units, the version of the metric system agreed upon in 1960 at the 11th General Conference on Weights and Measures. In metric nations it replaces the MKSA (meter-kilogram-second-ampere) system and in English-speaking nonmetric nations it replaces the FPS (foot-pound-second) system. The basic units of the SI system, as redefined in 1967, are the meter (length), kilogram (mass), second (time), ampere (electric current), kelvin (temperature) and candela (luminous intensity). Note that temperatures for most everyday uses are converted from kelvin to degrees Celsius. In 1971, the mole was added as the basic unit for the amount of a substance.

The SI system also has units derived from the basic units. The derived units include the newton (force), joule (energy or work), the watt (power), the coulomb (electric quantity), volt (electric potential difference and electromotive force), ohm (electric resistance), the henry (electric inductance), the weber (magnetic flux), the lumen (luminous flux), the pascal (pressure) and the siemens (electric conductance).

Although the above are the "official" basic and derived units of the SI system, a number of other metric units are acceptable including the liter and the standard atmosphere. Obsolete units include the kilogram-force, dyne, bar, technical and physical atmospheres torr, erg, the technical calorie and the metric horsepower.

**scruple** A unit of weight in the apothecaries' system. Originally, a small, sharp stone. Equal to 20 grains or ⅓ dram or 1.295 978 grams.

**second** The basic unit of time in the SI system defined, since 1967, as equal to 9 192 631 770 periods of radiation corresponding to the transition between the two hyperfine levels of the ground state of the atom of cesium-133. This atomic definition replaced the older definition of the ephemeris second which was an exact fraction (1/31 556 925.974 7) of the year 1900.

**siemens** Unit of electrical conductance, defined as the conductance in a conductor when a potential difference of one volt produces a current of one ampere. Also known as a mho or a reciprocal ohm.

**square centimeter** Unit of area equal to 0.000 1 square meter or 0.155 000 square inch.

**square foot** Unit of area equal to 144 square inches or 0.092 903 square meter (929 square centimeters).

**square inch** Unit of area equal to 6.451 600 square centimeters (0.000 645 square meter).

**square kilometer** Unit of area equal to 0.386 110 square miles or 247.105 4 acres.

**square meter** Unit of area equal to 10.763 910 square feet or 1.195 985 square yards.

**square mile** Unit of area equal to 640 acres or 2.589 988 square kilometers.

**square millimeter** Unit of area equal to 0.000 001 square meter or 0.001 549 9 square inch.

**square rod** Unit of area equal to 0.006 250 acre or 25.293 square meters.

**square yard** Unit of area equal to 0.836 127 square meters.

**stat-** prefix used in stating electrical units in the centimeter-gram-second system instead of the SI system.

**statampere** Unit of electric current equal to 0.000 000 000 333 564 ampere.

**statcoulomb** Unit of electric charge equal to 0.000 000 000 333 564 coulomb.

**statfarad** Unit of capacitance equal to 0.000 000 000 001 112 650 farad.

**stathenry** Unit of inductance equal to 898 755 400 000 henrys.

**statohm** Unit of electrical resistance equal to 898 755 400 000 ohms.

**statvolt** Unit of electrical force equal to 299.792 500 volts

**statweber** Unit of magnetic flux equal to 299.792 500 webers.

**steradian** A solid angle at the center of a sphere that subtends a spherical surface area equal to the square of the raidus of the sphere.

**stere** Unit of volume equal to one cubic meter or 1.31 cubic yards, usually used in measuring timber.

**stokes** A unit of kinematic viscosity of a fluid with a dynamic viscosity of one poise and a density of one gram per cubic centimeter; equal to 0.000 100 square meter per second or 0.001 076 391 square foot per second or 0.36 square meter per hour or 3.875 008 square feet per hour.

## T

**tablespoon** Unit of volume equal to 14.786 800 milliliters. Sixteen customary tablespoons equal one customary cup but it will require almost 17 customary tablespoons to equal one metric cup of 250 milliliters. It is likely the metric tablespoon will be defined as 15 milliliters.

**teaspoon** Unit of volume equal to 4.928 9 milliliters. It is likely that the metric teaspoon will be defined as 5 milliliters.

**ton** A unit of weight equal to 20 hundredweight; A short ton of 2 000 pounds is equal to 907.184 700 kilograms. A long ton of 2 240 pounds is equal to 101.604 7 kilograms. A metric ton is equal to 1 000 kilograms or 2 204.622 pounds.

**ton-force** A unit of force. A FPS ton-force is equal to 8 896.443 newtons or 0.907 185 metric ton-force. A metric ton-force is equal to 9 806.650 newtons or 1.102 311 FPS ton-force.

## W

**watt** SI unit of power defined as the power which produces energy at the rate of 1 joule per second, equal to 0.101 972 meter-kilogram per second or 0.737 562 foot-pound per second.

**weber** The basic unit of magnetic flux defined as the magnetic flux which, linking a circuit of one turn, produces an electromotive force of one volt as the flux is reduced to zero in one second. A weber equals one volt-second or 0.003 335 640 statweber or 100 000 000 maxwells. A weber also equals one telsa times one square meter or one joule per ampere.

## Y

**yard** A unit of length, originally a stick or spear, equal to three feet or 0.914 400 meter.

**yard per second** Unit of velocity equal to 0.914 400 meter per second.

**year** Length of time for the earth to make one revolution of the sun. The calendar or solar year is 365.242 198 790 mean solar days.

# DECIMAL EQUIVALENTS OF COMMON FRACTIONS

| Fraction | Decimal |
|---|---|
| 1/2 | 0.5 |
| 1/3 | 0.33333 |
| 2/3 | 0.66667 |
| 1/4 | 0.25 |
| 2/4 | 0.5 |
| 3/4 | 0.75 |
| 1/5 | 0.2 |
| 2/5 | 0.4 |
| 3/5 | 0.6 |
| 4/5 | 0.8 |
| 1/6 | 0.16667 |
| 2/6 | 0.33333 |
| 3/6 | 0.5 |
| 4/6 | 0.66667 |
| 5/6 | 0.83333 |
| 1/7 | 0.14286 |
| 2/7 | 0.28571 |
| 3/7 | 0.42857 |
| 4/7 | 0.57143 |
| 5/7 | 0.71429 |
| 6/7 | 0.85714 |
| 1/8 | 0.125 |
| 2/8 | 0.25 |
| 3/8 | 0.375 |
| 4/8 | 0.5 |
| 5/8 | 0.625 |
| 6/8 | 0.75 |
| 7/8 | 0.875 |
| 1/9 | 0.11111 |
| 2/9 | 0.22222 |
| 3/9 | 0.33333 |
| 4/9 | 0.44444 |
| 5/9 | 0.55555 |
| 6/9 | 0.66667 |
| 7/9 | 0.77778 |
| 8/9 | 0.88889 |
| 1/12 | 0.08333 |
| 2/12 | 0.16667 |
| 3/12 | 0.25 |
| 4/12 | 0.33333 |
| 5/12 | 0.41667 |
| 6/12 | 0.5 |
| 7/12 | 0.58333 |
| 8/12 | 0.66667 |
| 9/12 | 0.75 |
| 10/12 | 0.83333 |
| 11/12 | 0.91667 |
| 1/16 | 0.0625 |
| 2/16 | 0.125 |
| 3/16 | 0.1875 |
| 4/16 | 0.25 |
| 5/16 | 0.3125 |
| 6/16 | 0.375 |
| 7/16 | 0.4375 |
| 8/16 | 0.5 |
| 9/16 | 0.5625 |
| 10/16 | 0.625 |
| 11/16 | 0.6875 |
| 12/16 | 0.75 |
| 13/16 | 0.8125 |
| 14/16 | 0.875 |
| 15/16 | 0.9375 |
| 1/32 | 0.031250 |
| 2/32 | 0.0625 |
| 3/32 | 0.09375 |
| 4/32 | 0.125 |
| 5/32 | 0.15625 |
| 6/32 | 0.18750 |
| 7/32 | 0.21875 |
| 8/32 | 0.25 |
| 9/32 | 0.28125 |
| 10/32 | 0.3125 |
| 11/32 | 0.34375 |
| 12/32 | 0.375 |
| 13/32 | 0.40625 |
| 14/32 | 0.4375 |
| 15/32 | 0.46875 |
| 16/32 | 0.5 |
| 17/32 | 0.53125 |
| 18/32 | 0.5625 |
| 19/32 | 0.59375 |
| 20/32 | 0.625 |
| 21/32 | 0.65625 |
| 22/32 | 0.6875 |
| 23/32 | 0.71875 |
| 24/32 | 0.75 |

25/32 0.78125
26/32 0.8125
27/32 0.84375
28/32 0.875
29/32 0.90625
30/32 0.9375
31/32 0.96875

1/64 0.015625
2/64 0.03125
3/64 0.046875
4/64 0.0625
5/64 0.078125
6/64 0.093750
7/64 0.109375
8/64 0.125
9/64 0.140625
10/64 0.15625
11/64 0.171875
12/64 0.1875
13/64 0.203125
14/64 0.21875
15/64 0.234375
16/64 0.25
17/64 0.265625
18/64 0.28125
19/64 0.296875
20/64 0.3125
21/64 0.328125
22/64 0.34375
23/64 0.359375
24/64 0.375
25/64 0.390625
26/64 0.40625
27/64 0.421875
28/64 0.4375
29/64 0.453125
30/64 0.46875
31/64 0.484375
32/64 0.5
33/64 0.515625
34/64 0.53125
35/64 0.546875
36/64 0.5625
37/64 0.578125
38/64 0.59375
39/64 0.609375
40/64 0.625
41/64 0.640625
42/64 0.65625
43/64 0.671875
44/64 0.6875
45/64 0.703125
46/64 0.71875
47/64 0.734375
48/64 0.75
49/64 0.765625
50/64 0.78125
51/64 0.796875
52/64 0.8125
53/64 0.828125
54/64 0.84375
55/64 0.859375
56/64 0.875
57/64 0.890625
58/64 0.90625
59/64 0.921875
60/64 0.9375
61/64 0.953125
62/64 0.96875
63/64 0.984375

# Conversion Tables

## LENGTH
### Inches to Centimeters

| in | cm | in | cm |
|---|---|---|---|
| 1 | 2.540000 | 51 | 129.540000 |
| 2 | 5.080000 | 52 | 132.080000 |
| 3 | 7.620000 | 53 | 134.620000 |
| 4 | 10.160000 | 54 | 137.160000 |
| 5 | 12.700000 | 55 | 139.700000 |
| 6 | 15.240000 | 56 | 142.240000 |
| 7 | 17.780000 | 57 | 144.780000 |
| 8 | 20.320000 | 58 | 147.320000 |
| 9 | 22.860000 | 59 | 149.860000 |
| 10 | 25.400000 | 60 | 152.400000 |
| 11 | 27.940000 | 61 | 154.940000 |
| 12 | 30.480000 | 62 | 157.480000 |
| 13 | 33.020000 | 63 | 160.020000 |
| 14 | 35.560000 | 64 | 162.560000 |
| 15 | 38.100000 | 65 | 165.100000 |
| 16 | 40.640000 | 66 | 167.640000 |
| 17 | 43.180000 | 67 | 170.180000 |
| 18 | 45.720000 | 68 | 172.720000 |
| 19 | 48.260000 | 69 | 175.260000 |
| 20 | 50.800000 | 70 | 177.800000 |
| 21 | 53.340000 | 71 | 180.340000 |
| 22 | 55.880000 | 72 | 182.880000 |
| 23 | 58.420000 | 73 | 185.420000 |
| 24 | 60.960000 | 74 | 187.960000 |
| 25 | 63.500000 | 75 | 190.500000 |
| 26 | 66.040000 | 76 | 193.040000 |
| 27 | 68.580000 | 77 | 195.580000 |
| 28 | 71.120000 | 78 | 198.120000 |
| 29 | 73.660000 | 79 | 200.660000 |
| 30 | 76.200000 | 80 | 203.200000 |
| 31 | 78.740000 | 81 | 205.740000 |
| 32 | 81.280000 | 82 | 208.280000 |
| 33 | 83.820000 | 83 | 210.820000 |
| 34 | 86.360000 | 84 | 213.360000 |
| 35 | 88.900000 | 85 | 215.900000 |
| 36 | 91.440000 | 86 | 218.440000 |
| 37 | 93.980000 | 87 | 220.980000 |
| 38 | 96.520000 | 88 | 223.520000 |
| 39 | 99.060000 | 89 | 226.060000 |
| 40 | 101.600000 | 90 | 228.600000 |
| 41 | 104.140000 | 91 | 231.140000 |
| 42 | 106.680000 | 92 | 233.680000 |
| 43 | 109.220000 | 93 | 236.220000 |
| 44 | 111.760000 | 94 | 238.760000 |
| 45 | 114.300000 | 95 | 241.300000 |
| 46 | 116.840000 | 96 | 243.840000 |
| 47 | 119.380000 | 97 | 246.380000 |
| 48 | 121.920000 | 98 | 248.920000 |
| 49 | 124.460000 | 99 | 251.460000 |
| 50 | 127.000000 | 100 | 254.000000 |

## LENGTH
### Centimeters to Inches

| cm | in | cm | in |
|---|---|---|---|
| 1 | .393701 | 51 | 20.078751 |
| 2 | .787402 | 52 | 20.472452 |
| 3 | 1.181103 | 53 | 20.866153 |
| 4 | 1.574804 | 54 | 21.259854 |
| 5 | 1.968505 | 55 | 21.653555 |
| 6 | 2.362206 | 56 | 22.047256 |
| 7 | 2.755907 | 57 | 22.440957 |
| 8 | 3.149608 | 58 | 22.834658 |
| 9 | 3.543309 | 59 | 23.228359 |
| 10 | 3.937010 | 60 | 23.622060 |
| 11 | 4.330711 | 61 | 24.015761 |
| 12 | 4.724412 | 62 | 24.409462 |
| 13 | 5.118113 | 63 | 24.803163 |
| 14 | 5.511814 | 64 | 25.196864 |
| 15 | 5.905515 | 65 | 25.590565 |
| 16 | 6.299216 | 66 | 25.984266 |
| 17 | 6.692917 | 67 | 26.377967 |
| 18 | 7.086618 | 68 | 26.771668 |
| 19 | 7.480319 | 69 | 27.165369 |
| 20 | 7.874020 | 70 | 27.559070 |
| 21 | 8.267721 | 71 | 27.952771 |
| 22 | 8.661422 | 72 | 28.346472 |
| 23 | 9.055123 | 73 | 28.740173 |
| 24 | 9.448824 | 74 | 29.133874 |
| 25 | 9.842525 | 75 | 29.527575 |
| 26 | 10.236226 | 76 | 29.921276 |
| 27 | 10.629927 | 77 | 30.314977 |
| 28 | 11.023628 | 78 | 30.708678 |
| 29 | 11.417329 | 79 | 31.102379 |
| 30 | 11.811030 | 80 | 31.496080 |
| 31 | 12.204731 | 81 | 31.889781 |
| 32 | 12.598432 | 82 | 32.283482 |
| 33 | 12.992133 | 83 | 32.677183 |
| 34 | 13.385834 | 84 | 33.070884 |
| 35 | 13.779535 | 85 | 33.464585 |
| 36 | 14.173236 | 86 | 33.858286 |
| 37 | 14.566937 | 87 | 34.251987 |
| 38 | 14.960638 | 88 | 34.645688 |
| 39 | 15.354339 | 89 | 35.039389 |
| 40 | 15.748040 | 90 | 35.433090 |
| 41 | 16.141741 | 91 | 35.826791 |
| 42 | 16.535442 | 92 | 36.220492 |
| 43 | 16.929143 | 93 | 36.614193 |
| 44 | 17.322844 | 94 | 37.007894 |
| 45 | 17.716545 | 95 | 37.401595 |
| 46 | 18.110246 | 96 | 37.795296 |
| 47 | 18.503947 | 97 | 38.188997 |
| 48 | 18.897648 | 98 | 38.582698 |
| 49 | 19.291349 | 99 | 38.976399 |
| 50 | 19.685050 | 100 | 39.370100 |

## LENGTH
### Feet to Centimeters

| ft | cm | ft | cm |
|---|---|---|---|
| 1 | 30.480000 | 51 | 1554.480000 |
| 2 | 60.960000 | 52 | 1584.960000 |
| 3 | 91.440000 | 53 | 1615.440000 |
| 4 | 121.920000 | 54 | 1645.920000 |
| 5 | 152.400000 | 55 | 1676.400000 |
| 6 | 182.880000 | 56 | 1706.880000 |
| 7 | 213.360000 | 57 | 1737.360000 |
| 8 | 243.840000 | 58 | 1767.840000 |
| 9 | 274.320000 | 59 | 1798.320000 |
| 10 | 304.800000 | 60 | 1828.800000 |
| 11 | 335.280000 | 61 | 1859.280000 |
| 12 | 365.760000 | 62 | 1889.760000 |
| 13 | 396.240000 | 63 | 1920.240000 |
| 14 | 426.720000 | 64 | 1950.720000 |
| 15 | 457.200000 | 65 | 1981.200000 |
| 16 | 487.680000 | 66 | 2011.680000 |
| 17 | 518.160000 | 67 | 2042.160000 |
| 18 | 548.640000 | 68 | 2072.640000 |
| 19 | 579.120000 | 69 | 2103.120000 |
| 20 | 609.600000 | 70 | 2133.600000 |
| 21 | 640.080000 | 71 | 2164.080000 |
| 22 | 670.560000 | 72 | 2194.560000 |
| 23 | 701.040000 | 73 | 2225.040000 |
| 24 | 731.520000 | 74 | 2255.520000 |
| 25 | 762.000000 | 75 | 2286.000000 |
| 26 | 792.480000 | 76 | 2316.480000 |
| 27 | 822.960000 | 77 | 2346.960000 |
| 28 | 853.440000 | 78 | 2377.440000 |
| 29 | 883.920000 | 79 | 2407.920000 |
| 30 | 914.400000 | 80 | 2438.400000 |
| 31 | 944.880000 | 81 | 2468.880000 |
| 32 | 975.360000 | 82 | 2499.360000 |
| 33 | 1005.840000 | 83 | 2529.840000 |
| 34 | 1036.320000 | 84 | 2560.320000 |
| 35 | 1066.800000 | 85 | 2590.800000 |
| 36 | 1097.280000 | 86 | 2621.280000 |
| 37 | 1127.760000 | 87 | 2651.760000 |
| 38 | 1158.240000 | 88 | 2682.240000 |
| 39 | 1188.720000 | 89 | 2712.720000 |
| 40 | 1219.200000 | 90 | 2743.200000 |
| 41 | 1249.680000 | 91 | 2773.680000 |
| 42 | 1280.160000 | 92 | 2804.160000 |
| 43 | 1310.640000 | 93 | 2834.640000 |
| 44 | 1341.120000 | 94 | 2865.120000 |
| 45 | 1371.600000 | 95 | 2895.600000 |
| 46 | 1402.080000 | 96 | 2926.080000 |
| 47 | 1432.560000 | 97 | 2956.560000 |
| 48 | 1463.040000 | 98 | 2987.040000 |
| 49 | 1493.520000 | 99 | 3017.520000 |
| 50 | 1524.000000 | 100 | 3048.000000 |

## LENGTH
### Centimeters to Feet

| cm | ft | cm | ft |
|---|---|---|---|
| 1 | .032808 | 51 | 1.673208 |
| 2 | .065616 | 52 | 1.706016 |
| 3 | .098424 | 53 | 1.738824 |
| 4 | .131232 | 54 | 1.771632 |
| 5 | .164040 | 55 | 1.804440 |
| 6 | .196848 | 56 | 1.837248 |
| 7 | .229656 | 57 | 1.870056 |
| 8 | .262464 | 58 | 1.902864 |
| 9 | .295272 | 59 | 1.935672 |
| 10 | .328080 | 60 | 1.968480 |
| 11 | .360888 | 61 | 2.001288 |
| 12 | .393696 | 62 | 2.034096 |
| 13 | .426504 | 63 | 2.066904 |
| 14 | .459312 | 64 | 2.099712 |
| 15 | .492120 | 65 | 2.132520 |
| 16 | .524928 | 66 | 2.165328 |
| 17 | .557736 | 67 | 2.198136 |
| 18 | .590544 | 68 | 2.230944 |
| 19 | .623352 | 69 | 2.263752 |
| 20 | .656160 | 70 | 2.296560 |
| 21 | .688968 | 71 | 2.329368 |
| 22 | .721776 | 72 | 2.362176 |
| 23 | .754584 | 73 | 2.394984 |
| 24 | .787392 | 74 | 2.427792 |
| 25 | .820200 | 75 | 2.460600 |
| 26 | .853008 | 76 | 2.493408 |
| 27 | .885816 | 77 | 2.526216 |
| 28 | .918624 | 78 | 2.559024 |
| 29 | .951432 | 79 | 2.591832 |
| 30 | .984240 | 80 | 2.624640 |
| 31 | 1.017048 | 81 | 2.657448 |
| 32 | 1.049856 | 82 | 2.690256 |
| 33 | 1.082664 | 83 | 2.723064 |
| 34 | 1.115472 | 84 | 2.755872 |
| 35 | 1.148280 | 85 | 2.788680 |
| 36 | 1.181088 | 86 | 2.821488 |
| 37 | 1.213896 | 87 | 2.854296 |
| 38 | 1.246704 | 88 | 2.887104 |
| 39 | 1.279512 | 89 | 2.919912 |
| 40 | 1.312320 | 90 | 2.952720 |
| 41 | 1.345128 | 91 | 2.985528 |
| 42 | 1.377936 | 92 | 3.018336 |
| 43 | 1.410744 | 93 | 3.051144 |
| 44 | 1.443552 | 94 | 3.083952 |
| 45 | 1.476360 | 95 | 3.116760 |
| 46 | 1.509168 | 96 | 3.149568 |
| 47 | 1.541976 | 97 | 3.182376 |
| 48 | 1.574784 | 98 | 3.215184 |
| 49 | 1.607592 | 99 | 3.247992 |
| 50 | 1.640400 | 100 | 3.280800 |

## LENGTH
### Feet to Meters

| ft | m | ft | m |
|---|---|---|---|
| 1 | .304800 | 51 | 15.544800 |
| 2 | .609600 | 52 | 15.849600 |
| 3 | .914400 | 53 | 16.154400 |
| 4 | 1.219200 | 54 | 16.459200 |
| 5 | 1.524000 | 55 | 16.764000 |
| 6 | 1.828800 | 56 | 17.068800 |
| 7 | 2.133600 | 57 | 17.373600 |
| 8 | 2.438400 | 58 | 17.678400 |
| 9 | 2.743200 | 59 | 17.983200 |
| 10 | 3.048000 | 60 | 18.288000 |
| 11 | 3.352800 | 61 | 18.592800 |
| 12 | 3.657600 | 62 | 18.897600 |
| 13 | 3.962400 | 63 | 19.202400 |
| 14 | 4.267200 | 64 | 19.507200 |
| 15 | 4.572000 | 65 | 19.812000 |
| 16 | 4.876800 | 66 | 20.116800 |
| 17 | 5.181600 | 67 | 20.421600 |
| 18 | 5.486400 | 68 | 20.726400 |
| 19 | 5.791200 | 69 | 21.031200 |
| 20 | 6.096000 | 70 | 21.336000 |
| 21 | 6.400800 | 71 | 21.640800 |
| 22 | 6.705600 | 72 | 21.945600 |
| 23 | 7.010400 | 73 | 22.250400 |
| 24 | 7.315200 | 74 | 22.555200 |
| 25 | 7.620000 | 75 | 22.860000 |
| 26 | 7.924800 | 76 | 23.164800 |
| 27 | 8.229600 | 77 | 23.469600 |
| 28 | 8.534400 | 78 | 23.774400 |
| 29 | 8.839200 | 79 | 24.079200 |
| 30 | 9.144000 | 80 | 24.384000 |
| 31 | 9.448800 | 81 | 24.688800 |
| 32 | 9.753600 | 82 | 24.993600 |
| 33 | 10.058400 | 83 | 25.298400 |
| 34 | 10.363200 | 84 | 25.603200 |
| 35 | 10.668000 | 85 | 25.908000 |
| 36 | 10.972800 | 86 | 26.212800 |
| 37 | 11.277600 | 87 | 26.517600 |
| 38 | 11.582400 | 88 | 26.822400 |
| 39 | 11.887200 | 89 | 27.127200 |
| 40 | 12.192000 | 90 | 27.432000 |
| 41 | 12.496800 | 91 | 27.736800 |
| 42 | 12.801600 | 92 | 28.041600 |
| 43 | 13.106400 | 93 | 28.346400 |
| 44 | 13.411200 | 94 | 28.651200 |
| 45 | 13.716000 | 95 | 28.956000 |
| 46 | 14.020800 | 96 | 29.260800 |
| 47 | 14.325600 | 97 | 29.565600 |
| 48 | 14.630400 | 98 | 29.870400 |
| 49 | 14.935200 | 99 | 30.175200 |
| 50 | 15.240000 | 100 | 30.480000 |

## LENGTH
### Meters to Feet

| m | ft | m | ft |
|---|---|---|---|
| 1 | 3.280839 | 51 | 167.322789 |
| 2 | 6.561678 | 52 | 170.603628 |
| 3 | 9.842517 | 53 | 173.884467 |
| 4 | 13.123356 | 54 | 177.165306 |
| 5 | 16.404195 | 55 | 180.446145 |
| 6 | 19.685034 | 56 | 183.726984 |
| 7 | 22.965873 | 57 | 187.007823 |
| 8 | 26.246712 | 58 | 190.288662 |
| 9 | 29.527551 | 59 | 193.569501 |
| 10 | 32.808390 | 60 | 196.850340 |
| 11 | 36.089229 | 61 | 200.131179 |
| 12 | 39.370068 | 62 | 203.412018 |
| 13 | 42.650907 | 63 | 206.692857 |
| 14 | 45.931746 | 64 | 209.973696 |
| 15 | 49.212585 | 65 | 213.254535 |
| 16 | 52.493424 | 66 | 216.535374 |
| 17 | 55.774263 | 67 | 219.816213 |
| 18 | 59.055102 | 68 | 223.097052 |
| 19 | 62.335941 | 69 | 226.377891 |
| 20 | 65.616780 | 70 | 229.658730 |
| 21 | 68.897619 | 71 | 232.939569 |
| 22 | 72.178458 | 72 | 236.220408 |
| 23 | 75.459297 | 73 | 239.501247 |
| 24 | 78.740136 | 74 | 242.782086 |
| 25 | 82.020975 | 75 | 246.062925 |
| 26 | 85.301814 | 76 | 249.343764 |
| 27 | 88.582653 | 77 | 252.624603 |
| 28 | 91.863492 | 78 | 255.905442 |
| 29 | 95.144331 | 79 | 259.186281 |
| 30 | 98.425170 | 80 | 262.467120 |
| 31 | 101.706009 | 81 | 265.747959 |
| 32 | 104.986848 | 82 | 269.028798 |
| 33 | 108.267687 | 83 | 272.309637 |
| 34 | 111.548526 | 84 | 275.590476 |
| 35 | 114.829365 | 85 | 278.871315 |
| 36 | 118.110204 | 86 | 282.152154 |
| 37 | 121.391043 | 87 | 285.432993 |
| 38 | 124.671882 | 88 | 288.713832 |
| 39 | 127.952721 | 89 | 291.994671 |
| 40 | 131.233560 | 90 | 295.275510 |
| 41 | 134.514399 | 91 | 298.556349 |
| 42 | 137.795238 | 92 | 301.837188 |
| 43 | 141.076077 | 93 | 305.118027 |
| 44 | 144.356916 | 94 | 308.398866 |
| 45 | 147.637755 | 95 | 311.679705 |
| 46 | 150.918594 | 96 | 314.960544 |
| 47 | 154.199433 | 97 | 318.241383 |
| 48 | 157.480272 | 98 | 321.522222 |
| 49 | 160.761111 | 99 | 324.803061 |
| 50 | 164.041950 | 100 | 328.083900 |

## LENGTH
### Yards to Meters

| yd | m | yd | m |
|---|---|---|---|
| 1 | .914400 | 51 | 46.634400 |
| 2 | 1.828800 | 52 | 47.548800 |
| 3 | 2.743200 | 53 | 48.463200 |
| 4 | 3.657600 | 54 | 49.377600 |
| 5 | 4.572000 | 55 | 50.292000 |
| 6 | 5.486400 | 56 | 51.206400 |
| 7 | 6.400800 | 57 | 52.120800 |
| 8 | 7.315200 | 58 | 53.035200 |
| 9 | 8.229600 | 59 | 53.949600 |
| 10 | 9.144000 | 60 | 54.864000 |
| 11 | 10.058400 | 61 | 55.778400 |
| 12 | 10.972800 | 62 | 56.692800 |
| 13 | 11.887200 | 63 | 57.607200 |
| 14 | 12.801600 | 64 | 58.521600 |
| 15 | 13.716000 | 65 | 59.436000 |
| 16 | 14.630400 | 66 | 60.350400 |
| 17 | 15.544800 | 67 | 61.264800 |
| 18 | 16.459200 | 68 | 62.179200 |
| 19 | 17.373600 | 69 | 63.093600 |
| 20 | 18.288000 | 70 | 64.008000 |
| 21 | 19.202400 | 71 | 64.922400 |
| 22 | 20.116800 | 72 | 65.836800 |
| 23 | 21.031200 | 73 | 66.751200 |
| 24 | 21.945600 | 74 | 67.665600 |
| 25 | 22.860000 | 75 | 68.580000 |
| 26 | 23.774400 | 76 | 69.494400 |
| 27 | 24.688800 | 77 | 70.408800 |
| 28 | 25.603200 | 78 | 71.323200 |
| 29 | 26.517600 | 79 | 72.237600 |
| 30 | 27.432000 | 80 | 73.152000 |
| 31 | 28.346400 | 81 | 74.066400 |
| 32 | 29.260800 | 82 | 74.980800 |
| 33 | 30.175200 | 83 | 75.895200 |
| 34 | 31.089600 | 84 | 76.809600 |
| 35 | 32.004000 | 85 | 77.724000 |
| 36 | 32.918400 | 86 | 78.638400 |
| 37 | 33.832800 | 87 | 79.552800 |
| 38 | 34.747200 | 88 | 80.467200 |
| 39 | 35.661600 | 89 | 81.381600 |
| 40 | 36.576000 | 90 | 82.296000 |
| 41 | 37.490400 | 91 | 83.210400 |
| 42 | 38.404800 | 92 | 84.124800 |
| 43 | 39.319200 | 93 | 85.039200 |
| 44 | 40.233600 | 94 | 85.953600 |
| 45 | 41.148000 | 95 | 86.868000 |
| 46 | 42.062400 | 96 | 87.782400 |
| 47 | 42.976800 | 97 | 88.696800 |
| 48 | 43.891200 | 98 | 89.611200 |
| 49 | 44.805600 | 99 | 90.525600 |
| 50 | 45.720000 | 100 | 91.440000 |

## LENGTH
### Meters to Yards

| m | yd | m | yd |
|---|---|---|---|
| 1 | 1.093613 | 51 | 55.774263 |
| 2 | 2.187226 | 52 | 56.867876 |
| 3 | 3.280839 | 53 | 57.961489 |
| 4 | 4.374452 | 54 | 59.055102 |
| 5 | 5.468065 | 55 | 60.148715 |
| 6 | 6.561678 | 56 | 61.242328 |
| 7 | 7.655291 | 57 | 62.335941 |
| 8 | 8.748904 | 58 | 63.429554 |
| 9 | 9.842517 | 59 | 64.523167 |
| 10 | 10.936130 | 60 | 65.616780 |
| 11 | 12.029743 | 61 | 66.710393 |
| 12 | 13.123356 | 62 | 67.804006 |
| 13 | 14.216969 | 63 | 68.897619 |
| 14 | 15.310582 | 64 | 69.991232 |
| 15 | 16.404195 | 65 | 71.084845 |
| 16 | 17.497808 | 66 | 72.178458 |
| 17 | 18.591421 | 67 | 73.272071 |
| 18 | 19.685034 | 68 | 74.365684 |
| 19 | 20.778647 | 69 | 75.459297 |
| 20 | 21.872260 | 70 | 76.552910 |
| 21 | 22.965873 | 71 | 77.646523 |
| 22 | 24.059486 | 72 | 78.740136 |
| 23 | 25.153099 | 73 | 79.833749 |
| 24 | 26.246712 | 74 | 80.927362 |
| 25 | 27.340325 | 75 | 82.020975 |
| 26 | 28.433938 | 76 | 83.114588 |
| 27 | 29.527551 | 77 | 84.208201 |
| 28 | 30.621164 | 78 | 85.301814 |
| 29 | 31.714777 | 79 | 86.395427 |
| 30 | 32.808390 | 80 | 87.489040 |
| 31 | 33.902003 | 81 | 88.582653 |
| 32 | 34.995616 | 82 | 89.676266 |
| 33 | 36.089229 | 83 | 90.769879 |
| 34 | 37.182842 | 84 | 91.863492 |
| 35 | 38.276455 | 85 | 92.957105 |
| 36 | 39.370068 | 86 | 94.050718 |
| 37 | 40.463681 | 87 | 95.144331 |
| 38 | 41.557294 | 88 | 96.237944 |
| 39 | 42.650907 | 89 | 97.331557 |
| 40 | 43.744520 | 90 | 98.425170 |
| 41 | 44.838133 | 91 | 99.518783 |
| 42 | 45.931746 | 92 | 100.612396 |
| 43 | 47.025359 | 93 | 101.706009 |
| 44 | 48.118972 | 94 | 102.799622 |
| 45 | 49.212585 | 95 | 103.893235 |
| 46 | 50.306198 | 96 | 104.986848 |
| 47 | 51.399811 | 97 | 106.080461 |
| 48 | 52.493424 | 98 | 107.174074 |
| 49 | 53.587037 | 99 | 108.267687 |
| 50 | 54.680650 | 100 | 109.361300 |

# LENGTH
## Rods to Meters

| rd | m | rd | m |
|---|---|---|---|
| 1 | 5.029200 | 51 | 256.489200 |
| 2 | 10.058400 | 52 | 261.518400 |
| 3 | 15.087600 | 53 | 266.547600 |
| 4 | 20.116800 | 54 | 271.576800 |
| 5 | 25.146000 | 55 | 276.606000 |
| 6 | 30.175200 | 56 | 281.635200 |
| 7 | 35.204400 | 57 | 286.664400 |
| 8 | 40.233600 | 58 | 291.693600 |
| 9 | 45.262800 | 59 | 296.722800 |
| 10 | 50.292000 | 60 | 301.752000 |
| 11 | 55.321200 | 61 | 306.781200 |
| 12 | 60.350400 | 62 | 311.810400 |
| 13 | 65.379600 | 63 | 316.839600 |
| 14 | 70.408800 | 64 | 321.868800 |
| 15 | 75.438000 | 65 | 326.898000 |
| 16 | 80.467200 | 66 | 331.927200 |
| 17 | 85.496400 | 67 | 336.956400 |
| 18 | 90.525600 | 68 | 341.985600 |
| 19 | 95.554800 | 69 | 347.014800 |
| 20 | 100.584000 | 70 | 352.044000 |
| 21 | 105.613200 | 71 | 357.073200 |
| 22 | 110.642400 | 72 | 362.102400 |
| 23 | 115.671600 | 73 | 367.131600 |
| 24 | 120.700800 | 74 | 372.160800 |
| 25 | 125.730000 | 75 | 377.190000 |
| 26 | 130.759200 | 76 | 382.219200 |
| 27 | 135.788400 | 77 | 387.248400 |
| 28 | 140.817600 | 78 | 392.277600 |
| 29 | 145.846800 | 79 | 397.306800 |
| 30 | 150.876000 | 80 | 402.336000 |
| 31 | 155.905200 | 81 | 407.365200 |
| 32 | 160.934400 | 82 | 412.394400 |
| 33 | 165.963600 | 83 | 417.423600 |
| 34 | 170.992800 | 84 | 422.452800 |
| 35 | 176.022000 | 85 | 427.482000 |
| 36 | 181.051200 | 86 | 432.511200 |
| 37 | 186.080400 | 87 | 437.540400 |
| 38 | 191.109600 | 88 | 442.569600 |
| 39 | 196.138800 | 89 | 447.598800 |
| 40 | 201.168000 | 90 | 452.628000 |
| 41 | 206.197200 | 91 | 457.657200 |
| 42 | 211.226400 | 92 | 462.686400 |
| 43 | 216.255600 | 93 | 467.715600 |
| 44 | 221.284800 | 94 | 472.744800 |
| 45 | 226.314000 | 95 | 477.774000 |
| 46 | 231.343200 | 96 | 482.803200 |
| 47 | 236.372400 | 97 | 487.832400 |
| 48 | 241.401600 | 98 | 492.861600 |
| 49 | 246.430800 | 99 | 497.890800 |
| 50 | 251.460000 | 100 | 502.920000 |

# LENGTH
## Meters to Rods

| m | rd | m | rd |
|---|---|---|---|
| 1 | .198839 | 51 | 10.140789 |
| 2 | .397678 | 52 | 10.339628 |
| 3 | .596517 | 53 | 10.538467 |
| 4 | .795356 | 54 | 10.737306 |
| 5 | .994195 | 55 | 10.936145 |
| 6 | 1.193034 | 56 | 11.134984 |
| 7 | 1.391873 | 57 | 11.333823 |
| 8 | 1.590712 | 58 | 11.532662 |
| 9 | 1.789551 | 59 | 11.731501 |
| 10 | 1.988390 | 60 | 11.930340 |
| 11 | 2.187229 | 61 | 12.129179 |
| 12 | 2.386068 | 62 | 12.328018 |
| 13 | 2.584907 | 63 | 12.526857 |
| 14 | 2.783746 | 64 | 12.725696 |
| 15 | 2.982585 | 65 | 12.924535 |
| 16 | 3.181424 | 66 | 13.123374 |
| 17 | 3.380263 | 67 | 13.322213 |
| 18 | 3.579102 | 68 | 13.521052 |
| 19 | 3.777941 | 69 | 13.719891 |
| 20 | 3.976780 | 70 | 13.918730 |
| 21 | 4.175619 | 71 | 14.117569 |
| 22 | 4.374458 | 72 | 14.316408 |
| 23 | 4.573297 | 73 | 14.515247 |
| 24 | 4.772136 | 74 | 14.714086 |
| 25 | 4.970975 | 75 | 14.912925 |
| 26 | 5.169814 | 76 | 15.111764 |
| 27 | 5.368653 | 77 | 15.310603 |
| 28 | 5.567492 | 78 | 15.509442 |
| 29 | 5.766331 | 79 | 15.708281 |
| 30 | 5.965170 | 80 | 15.907120 |
| 31 | 6.164009 | 81 | 16.105959 |
| 32 | 6.362848 | 82 | 16.304798 |
| 33 | 6.561687 | 83 | 16.503637 |
| 34 | 6.760526 | 84 | 16.702476 |
| 35 | 6.959365 | 85 | 16.901315 |
| 36 | 7.158204 | 86 | 17.100154 |
| 37 | 7.357043 | 87 | 17.298993 |
| 38 | 7.555882 | 88 | 17.497832 |
| 39 | 7.754721 | 89 | 17.696671 |
| 40 | 7.953560 | 90 | 17.895510 |
| 41 | 8.152399 | 91 | 18.094349 |
| 42 | 8.351238 | 92 | 18.293188 |
| 43 | 8.550077 | 93 | 18.492027 |
| 44 | 8.748916 | 94 | 18.690866 |
| 45 | 8.947755 | 95 | 18.889705 |
| 46 | 9.146594 | 96 | 19.088544 |
| 47 | 9.345433 | 97 | 19.287383 |
| 48 | 9.544272 | 98 | 19.486222 |
| 49 | 9.743111 | 99 | 19.685061 |
| 50 | 9.941950 | 100 | 19.883900 |

## LENGTH
### Miles (stat) to Kilometers

| mi | km | mi | km |
|---|---|---|---|
| 1 | 1.609344 | 51 | 82.076544 |
| 2 | 3.218688 | 52 | 83.685888 |
| 3 | 4.828032 | 53 | 85.295232 |
| 4 | 6.437376 | 54 | 86.904576 |
| 5 | 8.046720 | 55 | 88.513920 |
| 6 | 9.656064 | 56 | 90.123264 |
| 7 | 11.265408 | 57 | 91.732608 |
| 8 | 12.874752 | 58 | 93.341952 |
| 9 | 14.484096 | 59 | 94.951296 |
| 10 | 16.093440 | 60 | 96.560640 |
| 11 | 17.702784 | 61 | 98.169984 |
| 12 | 19.312128 | 62 | 99.779328 |
| 13 | 20.921472 | 63 | 101.388672 |
| 14 | 22.530816 | 64 | 102.998016 |
| 15 | 24.140160 | 65 | 104.607360 |
| 16 | 25.749504 | 66 | 106.216704 |
| 17 | 27.358848 | 67 | 107.826048 |
| 18 | 28.968192 | 68 | 109.435392 |
| 19 | 30.577536 | 69 | 111.044736 |
| 20 | 32.186880 | 70 | 112.654080 |
| 21 | 33.796224 | 71 | 114.263424 |
| 22 | 35.405568 | 72 | 115.872768 |
| 23 | 37.014912 | 73 | 117.482112 |
| 24 | 38.624256 | 74 | 119.091456 |
| 25 | 40.233600 | 75 | 120.700800 |
| 26 | 41.842944 | 76 | 122.310144 |
| 27 | 43.452288 | 77 | 123.919488 |
| 28 | 45.061632 | 78 | 125.528832 |
| 29 | 46.670976 | 79 | 127.138176 |
| 30 | 48.280320 | 80 | 128.747520 |
| 31 | 49.889664 | 81 | 130.356864 |
| 32 | 51.499008 | 82 | 131.966208 |
| 33 | 53.108352 | 83 | 133.575552 |
| 34 | 54.717696 | 84 | 135.184896 |
| 35 | 56.327040 | 85 | 136.794240 |
| 36 | 57.936384 | 86 | 138.403584 |
| 37 | 59.545728 | 87 | 140.012928 |
| 38 | 61.155072 | 88 | 141.622272 |
| 39 | 62.764416 | 89 | 143.231616 |
| 40 | 64.373760 | 90 | 144.840960 |
| 41 | 65.983104 | 91 | 146.450304 |
| 42 | 67.592448 | 92 | 148.059648 |
| 43 | 69.201792 | 93 | 149.668992 |
| 44 | 70.811136 | 94 | 151.278336 |
| 45 | 72.420480 | 95 | 152.887680 |
| 46 | 74.029824 | 96 | 154.497024 |
| 47 | 75.639168 | 97 | 156.106368 |
| 48 | 77.248512 | 98 | 157.715712 |
| 49 | 78.857856 | 99 | 159.325056 |
| 50 | 80.467200 | 100 | 160.934400 |

## LENGTH
### Kilometers to Miles (stat)

| km | mi | km | mi |
|---|---|---|---|
| 1 | .621371 | 51 | 31.689921 |
| 2 | 1.242742 | 52 | 32.311292 |
| 3 | 1.864113 | 53 | 32.932663 |
| 4 | 2.485484 | 54 | 33.554034 |
| 5 | 3.106855 | 55 | 34.175405 |
| 6 | 3.728226 | 56 | 34.796776 |
| 7 | 4.349597 | 57 | 35.418147 |
| 8 | 4.970968 | 58 | 36.039518 |
| 9 | 5.592339 | 59 | 36.660889 |
| 10 | 6.213710 | 60 | 37.282260 |
| 11 | 6.835081 | 61 | 37.903631 |
| 12 | 7.456452 | 62 | 38.525002 |
| 13 | 8.077823 | 63 | 39.146373 |
| 14 | 8.699194 | 64 | 39.767744 |
| 15 | 9.320565 | 65 | 40.389115 |
| 16 | 9.941936 | 66 | 41.010486 |
| 17 | 10.563307 | 67 | 41.631857 |
| 18 | 11.184678 | 68 | 42.253228 |
| 19 | 11.806049 | 69 | 42.874599 |
| 20 | 12.427420 | 70 | 43.495970 |
| 21 | 13.048791 | 71 | 44.117341 |
| 22 | 13.670162 | 72 | 44.738712 |
| 23 | 14.291533 | 73 | 45.360083 |
| 24 | 14.912904 | 74 | 45.981454 |
| 25 | 15.534275 | 75 | 46.602825 |
| 26 | 16.155646 | 76 | 47.224196 |
| 27 | 16.777017 | 77 | 47.845567 |
| 28 | 17.398388 | 78 | 48.466938 |
| 29 | 18.019759 | 79 | 49.088309 |
| 30 | 18.641130 | 80 | 49.709680 |
| 31 | 19.262501 | 81 | 50.331051 |
| 32 | 19.883872 | 82 | 50.952422 |
| 33 | 20.505243 | 83 | 51.573793 |
| 34 | 21.126614 | 84 | 52.195164 |
| 35 | 21.747985 | 85 | 52.816535 |
| 36 | 22.369356 | 86 | 53.437906 |
| 37 | 22.990727 | 87 | 54.059277 |
| 38 | 23.612098 | 88 | 54.680648 |
| 39 | 24.233469 | 89 | 55.302019 |
| 40 | 24.854840 | 90 | 55.923390 |
| 41 | 25.476211 | 91 | 56.544761 |
| 42 | 26.097582 | 92 | 57.166132 |
| 43 | 26.718953 | 93 | 57.787503 |
| 44 | 27.340324 | 94 | 58.408874 |
| 45 | 27.961695 | 95 | 59.030245 |
| 46 | 28.583066 | 96 | 59.651616 |
| 47 | 29.204437 | 97 | 60.272987 |
| 48 | 29.825808 | 98 | 60.894358 |
| 49 | 30.447179 | 99 | 61.515729 |
| 50 | 31.068550 | 100 | 62.137100 |

## LENGTH
### Miles (naut) to Kilometers

| mi | km | mi | km |
|---|---|---|---|
| 1 | 1.852000 | 51 | 94.452000 |
| 2 | 3.704000 | 52 | 96.304000 |
| 3 | 5.556000 | 53 | 98.156000 |
| 4 | 7.408000 | 54 | 100.008000 |
| 5 | 9.260000 | 55 | 101.860000 |
| 6 | 11.112000 | 56 | 103.712000 |
| 7 | 12.964000 | 57 | 105.564000 |
| 8 | 14.816000 | 58 | 107.416000 |
| 9 | 16.668000 | 59 | 109.268000 |
| 10 | 18.520000 | 60 | 111.120000 |
| 11 | 20.372000 | 61 | 112.972000 |
| 12 | 22.224000 | 62 | 114.824000 |
| 13 | 24.076000 | 63 | 116.676000 |
| 14 | 25.928000 | 64 | 118.528000 |
| 15 | 27.780000 | 65 | 120.380000 |
| 16 | 29.632000 | 66 | 122.232000 |
| 17 | 31.484000 | 67 | 124.084000 |
| 18 | 33.336000 | 68 | 125.936000 |
| 19 | 35.188000 | 69 | 127.788000 |
| 20 | 37.040000 | 70 | 129.640000 |
| 21 | 38.892000 | 71 | 131.492000 |
| 22 | 40.744000 | 72 | 133.344000 |
| 23 | 42.596000 | 73 | 135.196000 |
| 24 | 44.448000 | 74 | 137.048000 |
| 25 | 46.300000 | 75 | 138.900000 |
| 26 | 48.152000 | 76 | 140.752000 |
| 27 | 50.004000 | 77 | 142.604000 |
| 28 | 51.856000 | 78 | 144.456000 |
| 29 | 53.708000 | 79 | 146.308000 |
| 30 | 55.560000 | 80 | 148.160000 |
| 31 | 57.412000 | 81 | 150.012000 |
| 32 | 59.264000 | 82 | 151.864000 |
| 33 | 61.116000 | 83 | 153.716000 |
| 34 | 62.968000 | 84 | 155.568000 |
| 35 | 64.820000 | 85 | 157.420000 |
| 36 | 66.672000 | 86 | 159.272000 |
| 37 | 68.524000 | 87 | 161.124000 |
| 38 | 70.376000 | 88 | 162.976000 |
| 39 | 72.228000 | 89 | 164.828000 |
| 40 | 74.080000 | 90 | 166.680000 |
| 41 | 75.932000 | 91 | 168.532000 |
| 42 | 77.784000 | 92 | 170.384000 |
| 43 | 79.636000 | 93 | 172.236000 |
| 44 | 81.488000 | 94 | 174.088000 |
| 45 | 83.340000 | 95 | 175.940000 |
| 46 | 85.192000 | 96 | 177.792000 |
| 47 | 87.044000 | 97 | 179.644000 |
| 48 | 88.896000 | 98 | 181.496000 |
| 49 | 90.748000 | 99 | 183.348000 |
| 50 | 92.600000 | 100 | 185.200000 |

# LENGTH
## Kilometers to Miles (naut)

| km | mi | km | mi |
|---|---|---|---|
| 1 | .539900 | 51 | 27.534900 |
| 2 | 1.079800 | 52 | 28.074800 |
| 3 | 1.619700 | 53 | 28.614700 |
| 4 | 2.159600 | 54 | 29.154600 |
| 5 | 2.699500 | 55 | 29.694500 |
| 6 | 3.239400 | 56 | 30.234400 |
| 7 | 3.779300 | 57 | 30.774300 |
| 8 | 4.319200 | 58 | 31.314200 |
| 9 | 4.859100 | 59 | 31.854100 |
| 10 | 5.399000 | 60 | 32.394000 |
| 11 | 5.938900 | 61 | 32.933900 |
| 12 | 6.478800 | 62 | 33.473800 |
| 13 | 7.018700 | 63 | 34.013700 |
| 14 | 7.558600 | 64 | 34.553600 |
| 15 | 8.098500 | 65 | 35.093500 |
| 16 | 8.638400 | 66 | 35.633400 |
| 17 | 9.178300 | 67 | 36.173300 |
| 18 | 9.718200 | 68 | 36.713200 |
| 19 | 10.258100 | 69 | 37.253100 |
| 20 | 10.798000 | 70 | 37.793000 |
| 21 | 11.337900 | 71 | 38.332900 |
| 22 | 11.877800 | 72 | 38.872800 |
| 23 | 12.417700 | 73 | 39.412700 |
| 24 | 12.957600 | 74 | 39.952600 |
| 25 | 13.497500 | 75 | 40.492500 |
| 26 | 14.037400 | 76 | 41.032400 |
| 27 | 14.577300 | 77 | 41.572300 |
| 28 | 15.117200 | 78 | 42.112200 |
| 29 | 15.657100 | 79 | 42.652100 |
| 30 | 16.197000 | 80 | 43.192000 |
| 31 | 16.736900 | 81 | 43.731900 |
| 32 | 17.276800 | 82 | 44.271800 |
| 33 | 17.816700 | 83 | 44.811700 |
| 34 | 18.356600 | 84 | 45.351600 |
| 35 | 18.896500 | 85 | 45.891500 |
| 36 | 19.436400 | 86 | 46.431400 |
| 37 | 19.976300 | 87 | 46.971300 |
| 38 | 20.516200 | 88 | 47.511200 |
| 39 | 21.056100 | 89 | 48.051100 |
| 40 | 21.596000 | 90 | 48.591000 |
| 41 | 22.135900 | 91 | 49.130900 |
| 42 | 22.675800 | 92 | 49.670800 |
| 43 | 23.215700 | 93 | 50.210700 |
| 44 | 23.755600 | 94 | 50.750600 |
| 45 | 24.295500 | 95 | 51.290500 |
| 46 | 24.835400 | 96 | 51.830400 |
| 47 | 25.375300 | 97 | 52.370300 |
| 48 | 25.915200 | 98 | 52.910200 |
| 49 | 26.455100 | 99 | 53.450100 |
| 50 | 26.995000 | 100 | 53.990000 |

## AREA
### Square Inches to Square Centimeters

| $in^2$ | $cm^2$ | $in^2$ | $cm^2$ |
|---|---|---|---|
| 1 | 6.451600 | 51 | 329.031600 |
| 2 | 12.903200 | 52 | 335.483200 |
| 3 | 19.354800 | 53 | 341.934800 |
| 4 | 25.806400 | 54 | 348.386400 |
| 5 | 32.258000 | 55 | 354.838000 |
| 6 | 38.709600 | 56 | 361.289600 |
| 7 | 45.161200 | 57 | 367.741200 |
| 8 | 51.612800 | 58 | 374.192800 |
| 9 | 58.064400 | 59 | 380.644400 |
| 10 | 64.516000 | 60 | 387.096000 |
| 11 | 70.967600 | 61 | 393.547600 |
| 12 | 77.419200 | 62 | 399.999200 |
| 13 | 83.870800 | 63 | 406.450800 |
| 14 | 90.322400 | 64 | 412.902400 |
| 15 | 96.774000 | 65 | 419.354000 |
| 16 | 103.225600 | 66 | 425.805600 |
| 17 | 109.677200 | 67 | 432.257200 |
| 18 | 116.128800 | 68 | 438.708800 |
| 19 | 122.580400 | 69 | 445.160400 |
| 20 | 129.032000 | 70 | 451.612000 |
| 21 | 135.483600 | 71 | 458.063600 |
| 22 | 141.935200 | 72 | 464.515200 |
| 23 | 148.386800 | 73 | 470.966800 |
| 24 | 154.838400 | 74 | 477.418400 |
| 25 | 161.290000 | 75 | 483.870000 |
| 26 | 167.741600 | 76 | 490.321600 |
| 27 | 174.193200 | 77 | 496.773200 |
| 28 | 180.644800 | 78 | 503.224800 |
| 29 | 187.096400 | 79 | 509.676400 |
| 30 | 193.548000 | 80 | 516.128000 |
| 31 | 199.999600 | 81 | 522.579600 |
| 32 | 206.451200 | 82 | 529.031200 |
| 33 | 212.902800 | 83 | 535.482800 |
| 34 | 219.354400 | 84 | 541.934400 |
| 35 | 225.806000 | 85 | 548.386000 |
| 36 | 232.257600 | 86 | 554.837600 |
| 37 | 238.709200 | 87 | 561.289200 |
| 38 | 245.160800 | 88 | 567.740800 |
| 39 | 251.612400 | 89 | 574.192400 |
| 40 | 258.064000 | 90 | 580.644000 |
| 41 | 264.515600 | 91 | 587.095600 |
| 42 | 270.967200 | 92 | 593.547200 |
| 43 | 277.418800 | 93 | 599.998800 |
| 44 | 283.870400 | 94 | 606.450400 |
| 45 | 290.322000 | 95 | 612.902000 |
| 46 | 296.773600 | 96 | 619.353600 |
| 47 | 303.225200 | 97 | 625.805200 |
| 48 | 309.676800 | 98 | 632.256800 |
| 49 | 316.128400 | 99 | 638.708400 |
| 50 | 322.580000 | 100 | 645.160000 |

# AREA
## Square Centimeters to Square Inches

| cm² | in² | cm² | in² |
|---|---|---|---|
| 1 | .155000 | 51 | 7.905000 |
| 2 | .310000 | 52 | 8.060000 |
| 3 | .465000 | 53 | 8.215000 |
| 4 | .620000 | 54 | 8.370000 |
| 5 | .775000 | 55 | 8.525000 |
| 6 | .930000 | 56 | 8.680000 |
| 7 | 1.085000 | 57 | 8.835000 |
| 8 | 1.240000 | 58 | 8.990000 |
| 9 | 1.395000 | 59 | 9.145000 |
| 10 | 1.550000 | 60 | 9.300000 |
| 11 | 1.705000 | 61 | 9.455000 |
| 12 | 1.860000 | 62 | 9.610000 |
| 13 | 2.015000 | 63 | 9.765000 |
| 14 | 2.170000 | 64 | 9.920000 |
| 15 | 2.325000 | 65 | 10.075000 |
| 16 | 2.480000 | 66 | 10.230000 |
| 17 | 2.635000 | 67 | 10.385000 |
| 18 | 2.790000 | 68 | 10.540000 |
| 19 | 2.945000 | 69 | 10.695000 |
| 20 | 3.100000 | 70 | 10.850000 |
| 21 | 3.255000 | 71 | 11.005000 |
| 22 | 3.410000 | 72 | 11.160000 |
| 23 | 3.565000 | 73 | 11.315000 |
| 24 | 3.720000 | 74 | 11.470000 |
| 25 | 3.875000 | 75 | 11.625000 |
| 26 | 4.030000 | 76 | 11.780000 |
| 27 | 4.185000 | 77 | 11.935000 |
| 28 | 4.340000 | 78 | 12.090000 |
| 29 | 4.495000 | 79 | 12.245000 |
| 30 | 4.650000 | 80 | 12.400000 |
| 31 | 4.805000 | 81 | 12.555000 |
| 32 | 4.960000 | 82 | 12.710000 |
| 33 | 5.115000 | 83 | 12.865000 |
| 34 | 5.270000 | 84 | 13.020000 |
| 35 | 5.425000 | 85 | 13.175000 |
| 36 | 5.580000 | 86 | 13.330000 |
| 37 | 5.735000 | 87 | 13.485000 |
| 38 | 5.890000 | 88 | 13.640000 |
| 39 | 6.045000 | 89 | 13.795000 |
| 40 | 6.200000 | 90 | 13.950000 |
| 41 | 6.355000 | 91 | 14.105000 |
| 42 | 6.510000 | 92 | 14.260000 |
| 43 | 6.665000 | 93 | 14.415000 |
| 44 | 6.820000 | 94 | 14.570000 |
| 45 | 6.975000 | 95 | 14.725000 |
| 46 | 7.130000 | 96 | 14.880000 |
| 47 | 7.285000 | 97 | 15.035000 |
| 48 | 7.440000 | 98 | 15.190000 |
| 49 | 7.595000 | 99 | 15.345000 |
| 50 | 7.750000 | 100 | 15.500000 |

## AREA
### Square Feet to Square Meters

| ft² | m² | ft² | m² |
|---|---|---|---|
| 1 | .092903 | 51 | 4.738053 |
| 2 | .185806 | 52 | 4.830956 |
| 3 | .278709 | 53 | 4.923859 |
| 4 | .371612 | 54 | 5.016762 |
| 5 | .464515 | 55 | 5.109665 |
| 6 | .557418 | 56 | 5.202568 |
| 7 | .650321 | 57 | 5.295471 |
| 8 | .743224 | 58 | 5.388374 |
| 9 | .836127 | 59 | 5.481277 |
| 10 | .929030 | 60 | 5.574180 |
| 11 | 1.021933 | 61 | 5.667083 |
| 12 | 1.114836 | 62 | 5.759986 |
| 13 | 1.207739 | 63 | 5.852889 |
| 14 | 1.300642 | 64 | 5.945792 |
| 15 | 1.393545 | 65 | 6.038695 |
| 16 | 1.486448 | 66 | 6.131598 |
| 17 | 1.579351 | 67 | 6.224501 |
| 18 | 1.672254 | 68 | 6.317404 |
| 19 | 1.765157 | 69 | 6.410307 |
| 20 | 1.858060 | 70 | 6.503210 |
| 21 | 1.950963 | 71 | 6.596113 |
| 22 | 2.043866 | 72 | 6.689016 |
| 23 | 2.136769 | 73 | 6.781919 |
| 24 | 2.229672 | 74 | 6.874822 |
| 25 | 2.322575 | 75 | 6.967725 |
| 26 | 2.415478 | 76 | 7.060628 |
| 27 | 2.508381 | 77 | 7.153531 |
| 28 | 2.601284 | 78 | 7.246434 |
| 29 | 2.694187 | 79 | 7.339337 |
| 30 | 2.787090 | 80 | 7.432240 |
| 31 | 2.879993 | 81 | 7.525143 |
| 32 | 2.972896 | 82 | 7.618046 |
| 33 | 3.065799 | 83 | 7.710949 |
| 34 | 3.158702 | 84 | 7.803852 |
| 35 | 3.251605 | 85 | 7.896755 |
| 36 | 3.344508 | 86 | 7.989658 |
| 37 | 3.437411 | 87 | 8.082561 |
| 38 | 3.530314 | 88 | 8.175464 |
| 39 | 3.623217 | 89 | 8.268367 |
| 40 | 3.716120 | 90 | 8.361270 |
| 41 | 3.809023 | 91 | 8.454173 |
| 42 | 3.901926 | 92 | 8.547076 |
| 43 | 3.994829 | 93 | 8.639979 |
| 44 | 4.087732 | 94 | 8.732882 |
| 45 | 4.180635 | 95 | 8.825785 |
| 46 | 4.273538 | 96 | 8.918688 |
| 47 | 4.366441 | 97 | 9.011591 |
| 48 | 4.459344 | 98 | 9.104494 |
| 49 | 4.552247 | 99 | 9.197397 |
| 50 | 4.645150 | 100 | 9.290300 |

## AREA
### Square Meters to Square Feet

| $m^2$ | $ft^2$ | $m^2$ | $ft^2$ |
|---|---|---|---|
| 1 | 10.763910 | 51 | 548.959410 |
| 2 | 21.527820 | 52 | 559.723320 |
| 3 | 32.291730 | 53 | 570.487230 |
| 4 | 43.055640 | 54 | 581.251140 |
| 5 | 53.819550 | 55 | 592.015050 |
| 6 | 64.583460 | 56 | 602.778960 |
| 7 | 75.347370 | 57 | 613.542870 |
| 8 | 86.111280 | 58 | 624.306780 |
| 9 | 96.875190 | 59 | 635.070690 |
| 10 | 107.639100 | 60 | 645.834600 |
| 11 | 118.403010 | 61 | 656.598510 |
| 12 | 129.166920 | 62 | 667.362420 |
| 13 | 139.930830 | 63 | 678.126330 |
| 14 | 150.694740 | 64 | 688.890240 |
| 15 | 161.458650 | 65 | 699.654150 |
| 16 | 172.222560 | 66 | 710.418060 |
| 17 | 182.986470 | 67 | 721.181970 |
| 18 | 193.750380 | 68 | 731.945880 |
| 19 | 204.514290 | 69 | 742.709790 |
| 20 | 215.278200 | 70 | 753.473700 |
| 21 | 226.042110 | 71 | 764.237610 |
| 22 | 236.806020 | 72 | 775.001520 |
| 23 | 247.569930 | 73 | 785.765430 |
| 24 | 258.333840 | 74 | 796.529340 |
| 25 | 269.097750 | 75 | 807.293250 |
| 26 | 279.861660 | 76 | 818.057160 |
| 27 | 290.625570 | 77 | 828.821070 |
| 28 | 301.389480 | 78 | 839.584980 |
| 29 | 312.153390 | 79 | 850.348890 |
| 30 | 322.917300 | 80 | 861.112800 |
| 31 | 333.681210 | 81 | 871.876710 |
| 32 | 344.445120 | 82 | 882.640620 |
| 33 | 355.209030 | 83 | 893.404530 |
| 34 | 365.972940 | 84 | 904.168440 |
| 35 | 376.736850 | 85 | 914.932350 |
| 36 | 387.500760 | 86 | 925.696260 |
| 37 | 398.264670 | 87 | 936.460170 |
| 38 | 409.028580 | 88 | 947.224080 |
| 39 | 419.792490 | 89 | 957.987990 |
| 40 | 430.556400 | 90 | 968.751900 |
| 41 | 441.320310 | 91 | 979.515810 |
| 42 | 452.084220 | 92 | 990.279720 |
| 43 | 462.848130 | 93 | 1001.043630 |
| 44 | 473.612040 | 94 | 1011.807540 |
| 45 | 484.375950 | 95 | 1022.571450 |
| 46 | 495.139860 | 96 | 1033.335360 |
| 47 | 505.903770 | 97 | 1044.099270 |
| 48 | 516.667680 | 98 | 1054.863180 |
| 49 | 527.431590 | 99 | 1065.627090 |
| 50 | 538.195500 | 100 | 1076.391000 |

# AREA
## Square Yards to Square Meters

| $yd^2$ | $m^2$ | $yd^2$ | $m^2$ |
|---|---|---|---|
| 1 | .836127 | 51 | 42.642477 |
| 2 | 1.672254 | 52 | 43.478604 |
| 3 | 2.508381 | 53 | 44.314731 |
| 4 | 3.344508 | 54 | 45.150858 |
| 5 | 4.180635 | 55 | 45.986985 |
| 6 | 5.016762 | 56 | 46.823112 |
| 7 | 5.852889 | 57 | 47.659239 |
| 8 | 6.689016 | 58 | 48.495366 |
| 9 | 7.525143 | 59 | 49.331493 |
| 10 | 8.361270 | 60 | 50.167620 |
| 11 | 9.197397 | 61 | 51.003747 |
| 12 | 10.033524 | 62 | 51.839874 |
| 13 | 10.869651 | 63 | 52.676001 |
| 14 | 11.705778 | 64 | 53.512128 |
| 15 | 12.541905 | 65 | 54.348255 |
| 16 | 13.378032 | 66 | 55.184382 |
| 17 | 14.214159 | 67 | 56.020509 |
| 18 | 15.050286 | 68 | 56.856636 |
| 19 | 15.886413 | 69 | 57.692763 |
| 20 | 16.722540 | 70 | 58.528890 |
| 21 | 17.558667 | 71 | 59.365017 |
| 22 | 18.394794 | 72 | 60.201144 |
| 23 | 19.230921 | 73 | 61.037271 |
| 24 | 20.067048 | 74 | 61.873398 |
| 25 | 20.903175 | 75 | 62.709525 |
| 26 | 21.739302 | 76 | 63.545652 |
| 27 | 22.575429 | 77 | 64.381779 |
| 28 | 23.411556 | 78 | 65.217906 |
| 29 | 24.247683 | 79 | 66.054033 |
| 30 | 25.083810 | 80 | 66.890160 |
| 31 | 25.919937 | 81 | 67.726287 |
| 32 | 26.756064 | 82 | 68.562414 |
| 33 | 27.592191 | 83 | 69.398541 |
| 34 | 28.428318 | 84 | 70.234668 |
| 35 | 29.264445 | 85 | 71.070795 |
| 36 | 30.100572 | 86 | 71.906922 |
| 37 | 30.936699 | 87 | 72.743049 |
| 38 | 31.772826 | 88 | 73.579176 |
| 39 | 32.608953 | 89 | 74.415303 |
| 40 | 33.445080 | 90 | 75.251430 |
| 41 | 34.281207 | 91 | 76.087557 |
| 42 | 35.117334 | 92 | 76.923684 |
| 43 | 35.953461 | 93 | 77.759811 |
| 44 | 36.789588 | 94 | 78.595938 |
| 45 | 37.625715 | 95 | 79.432065 |
| 46 | 38.461842 | 96 | 80.268192 |
| 47 | 39.297969 | 97 | 81.104319 |
| 48 | 40.134096 | 98 | 81.940446 |
| 49 | 40.970223 | 99 | 82.776573 |
| 50 | 41.806350 | 100 | 83.612700 |

## AREA
### Square Meters to Square Yards

| m² | yd² | m² | yd² |
|---|---|---|---|
| 1 | 1.195985 | 51 | 60.995235 |
| 2 | 2.391970 | 52 | 62.191220 |
| 3 | 3.587955 | 53 | 63.387205 |
| 4 | 4.783940 | 54 | 64.583190 |
| 5 | 5.979925 | 55 | 65.779175 |
| 6 | 7.175910 | 56 | 66.975160 |
| 7 | 8.371895 | 57 | 68.171145 |
| 8 | 9.567880 | 58 | 69.367130 |
| 9 | 10.763865 | 59 | 70.563115 |
| 10 | 11.959850 | 60 | 71.759100 |
| 11 | 13.155835 | 61 | 72.955085 |
| 12 | 14.351820 | 62 | 74.151070 |
| 13 | 15.547805 | 63 | 75.347055 |
| 14 | 16.743790 | 64 | 76.543040 |
| 15 | 17.939775 | 65 | 77.739025 |
| 16 | 19.135760 | 66 | 78.935010 |
| 17 | 20.331745 | 67 | 80.130995 |
| 18 | 21.527730 | 68 | 81.326980 |
| 19 | 22.723715 | 69 | 82.522965 |
| 20 | 23.919700 | 70 | 83.718950 |
| 21 | 25.115685 | 71 | 84.914935 |
| 22 | 26.311670 | 72 | 86.110920 |
| 23 | 27.507655 | 73 | 87.306905 |
| 24 | 28.703640 | 74 | 88.502890 |
| 25 | 29.899625 | 75 | 89.698875 |
| 26 | 31.095610 | 76 | 90.894860 |
| 27 | 32.291595 | 77 | 92.090845 |
| 28 | 33.487580 | 78 | 93.286830 |
| 29 | 34.683565 | 79 | 94.482815 |
| 30 | 35.879550 | 80 | 95.678800 |
| 31 | 37.075535 | 81 | 96.874785 |
| 32 | 38.271520 | 82 | 98.070770 |
| 33 | 39.467505 | 83 | 99.266755 |
| 34 | 40.663490 | 84 | 100.462740 |
| 35 | 41.859475 | 85 | 101.658725 |
| 36 | 43.055460 | 86 | 102.854710 |
| 37 | 44.251445 | 87 | 104.050695 |
| 38 | 45.447430 | 88 | 105.246680 |
| 39 | 46.643415 | 89 | 106.442665 |
| 40 | 47.839400 | 90 | 107.638650 |
| 41 | 49.035385 | 91 | 108.834635 |
| 42 | 50.231370 | 92 | 110.030620 |
| 43 | 51.427355 | 93 | 111.226605 |
| 44 | 52.623340 | 94 | 112.422590 |
| 45 | 53.819325 | 95 | 113.618575 |
| 46 | 55.015310 | 96 | 114.814560 |
| 47 | 56.211295 | 97 | 116.010545 |
| 48 | 57.407280 | 98 | 117.206530 |
| 49 | 58.603265 | 99 | 118.402515 |
| 50 | 59.799250 | 100 | 119.598500 |

## AREA
### Square Rods to Square Meters

| rd² | m² | rd² | m² |
|---|---|---|---|
| 1 | 25.293000 | 51 | 1289.943000 |
| 2 | 50.586000 | 52 | 1315.236000 |
| 3 | 75.879000 | 53 | 1340.529000 |
| 4 | 101.172000 | 54 | 1365.822000 |
| 5 | 126.465000 | 55 | 1391.115000 |
| 6 | 151.758000 | 56 | 1416.408000 |
| 7 | 177.051000 | 57 | 1441.701000 |
| 8 | 202.344000 | 58 | 1466.994000 |
| 9 | 227.637000 | 59 | 1492.287000 |
| 10 | 252.930000 | 60 | 1517.580000 |
| 11 | 278.223000 | 61 | 1542.873000 |
| 12 | 303.516000 | 62 | 1568.166000 |
| 13 | 328.809000 | 63 | 1593.459000 |
| 14 | 354.102000 | 64 | 1618.752000 |
| 15 | 379.395000 | 65 | 1644.045000 |
| 16 | 404.688000 | 66 | 1669.338000 |
| 17 | 429.981000 | 67 | 1694.631000 |
| 18 | 455.274000 | 68 | 1719.924000 |
| 19 | 480.567000 | 69 | 1745.217000 |
| 20 | 505.860000 | 70 | 1770.510000 |
| 21 | 531.153000 | 71 | 1795.803000 |
| 22 | 556.446000 | 72 | 1821.096000 |
| 23 | 581.739000 | 73 | 1846.389000 |
| 24 | 607.032000 | 74 | 1871.682000 |
| 25 | 632.325000 | 75 | 1896.975000 |
| 26 | 657.618000 | 76 | 1922.268000 |
| 27 | 682.911000 | 77 | 1947.561000 |
| 28 | 708.204000 | 78 | 1972.854000 |
| 29 | 733.497000 | 79 | 1998.147000 |
| 30 | 758.790000 | 80 | 2023.440000 |
| 31 | 784.083000 | 81 | 2048.733000 |
| 32 | 809.376000 | 82 | 2074.026000 |
| 33 | 834.669000 | 83 | 2099.319000 |
| 34 | 859.962000 | 84 | 2124.612000 |
| 35 | 885.255000 | 85 | 2149.905000 |
| 36 | 910.548000 | 86 | 2175.198000 |
| 37 | 935.841000 | 87 | 2200.491000 |
| 38 | 961.134000 | 88 | 2225.784000 |
| 39 | 986.427000 | 89 | 2251.077000 |
| 40 | 1011.720000 | 90 | 2276.370000 |
| 41 | 1037.013000 | 91 | 2301.663000 |
| 42 | 1062.306000 | 92 | 2326.956000 |
| 43 | 1087.599000 | 93 | 2352.249000 |
| 44 | 1112.892000 | 94 | 2377.542000 |
| 45 | 1138.185000 | 95 | 2402.835000 |
| 46 | 1163.478000 | 96 | 2428.128000 |
| 47 | 1188.771000 | 97 | 2453.421000 |
| 48 | 1214.064000 | 98 | 2478.714000 |
| 49 | 1239.357000 | 99 | 2504.007000 |
| 50 | 1264.650000 | 100 | 2529.300000 |

## AREA
### Square Meters to Square Rods

| m² | rd² | m² | rd² |
|---|---|---|---|
| 1 | .039540 | 51 | 2.016540 |
| 2 | .079080 | 52 | 2.056080 |
| 3 | .118620 | 53 | 2.095620 |
| 4 | .158160 | 54 | 2.135160 |
| 5 | .197700 | 55 | 2.174700 |
| 6 | .237240 | 56 | 2.214240 |
| 7 | .276780 | 57 | 2.253780 |
| 8 | .316320 | 58 | 2.293320 |
| 9 | .355860 | 59 | 2.332860 |
| 10 | .395400 | 60 | 2.372400 |
| 11 | .434940 | 61 | 2.411940 |
| 12 | .474480 | 62 | 2.451480 |
| 13 | .514020 | 63 | 2.491020 |
| 14 | .553560 | 64 | 2.530560 |
| 15 | .593100 | 65 | 2.570100 |
| 16 | .632640 | 66 | 2.609640 |
| 17 | .672180 | 67 | 2.649180 |
| 18 | .711720 | 68 | 2.688720 |
| 19 | .751260 | 69 | 2.728260 |
| 20 | .790800 | 70 | 2.767800 |
| 21 | .830340 | 71 | 2.807340 |
| 22 | .869880 | 72 | 2.846880 |
| 23 | .909420 | 73 | 2.886420 |
| 24 | .948960 | 74 | 2.925960 |
| 25 | .988500 | 75 | 2.965500 |
| 26 | 1.028040 | 76 | 3.005040 |
| 27 | 1.067580 | 77 | 3.044580 |
| 28 | 1.107120 | 78 | 3.084120 |
| 29 | 1.146660 | 79 | 3.123660 |
| 30 | 1.186200 | 80 | 3.163200 |
| 31 | 1.225740 | 81 | 3.202740 |
| 32 | 1.265280 | 82 | 3.242280 |
| 33 | 1.304820 | 83 | 3.281820 |
| 34 | 1.344360 | 84 | 3.321360 |
| 35 | 1.383900 | 85 | 3.360900 |
| 36 | 1.423440 | 86 | 3.400440 |
| 37 | 1.462980 | 87 | 3.439980 |
| 38 | 1.502520 | 88 | 3.479520 |
| 39 | 1.542060 | 89 | 3.519060 |
| 40 | 1.581600 | 90 | 3.558600 |
| 41 | 1.621140 | 91 | 3.598140 |
| 42 | 1.660680 | 92 | 3.637680 |
| 43 | 1.700220 | 93 | 3.677220 |
| 44 | 1.739760 | 94 | 3.716760 |
| 45 | 1.779300 | 95 | 3.756300 |
| 46 | 1.818840 | 96 | 3.795840 |
| 47 | 1.858380 | 97 | 3.835380 |
| 48 | 1.897920 | 98 | 3.874920 |
| 49 | 1.937460 | 99 | 3.914460 |
| 50 | 1.977000 | 100 | 3.954000 |

## AREA
### Acres to Square Meters

| a | m² | a | m² |
|---|---|---|---|
| 1 | 4046.856000 | 51 | 206389.656000 |
| 2 | 8093.712000 | 52 | 210436.512000 |
| 3 | 12140.568000 | 53 | 214483.368000 |
| 4 | 16187.424000 | 54 | 218530.224000 |
| 5 | 20234.280000 | 55 | 222577.080000 |
| 6 | 24281.136000 | 56 | 226623.936000 |
| 7 | 28327.992000 | 57 | 230670.792000 |
| 8 | 32374.848000 | 58 | 234717.648000 |
| 9 | 36421.704000 | 59 | 238764.504000 |
| 10 | 40468.560000 | 60 | 242811.360000 |
| 11 | 44515.416000 | 61 | 246858.216000 |
| 12 | 48562.272000 | 62 | 250905.072000 |
| 13 | 52609.128000 | 63 | 254951.928000 |
| 14 | 56655.984000 | 64 | 258998.784000 |
| 15 | 60702.840000 | 65 | 263045.640000 |
| 16 | 64749.696000 | 66 | 267092.496000 |
| 17 | 68796.552000 | 67 | 271139.352000 |
| 18 | 72843.408000 | 68 | 275186.208000 |
| 19 | 76890.264000 | 69 | 279233.064000 |
| 20 | 80937.120000 | 70 | 283279.920000 |
| 21 | 84983.976000 | 71 | 287326.776000 |
| 22 | 89030.832000 | 72 | 291373.632000 |
| 23 | 93077.688000 | 73 | 295420.488000 |
| 24 | 97124.544000 | 74 | 299467.344000 |
| 25 | 101171.400000 | 75 | 303514.200000 |
| 26 | 105218.256000 | 76 | 307561.056000 |
| 27 | 109265.112000 | 77 | 311607.912000 |
| 28 | 113311.968000 | 78 | 315654.768000 |
| 29 | 117358.824000 | 79 | 319701.624000 |
| 30 | 121405.680000 | 80 | 323748.480000 |
| 31 | 125452.536000 | 81 | 327795.336000 |
| 32 | 129499.392000 | 82 | 331842.192000 |
| 33 | 133546.248000 | 83 | 335889.048000 |
| 34 | 137593.104000 | 84 | 339935.904000 |
| 35 | 141639.960000 | 85 | 343982.760000 |
| 36 | 145686.816000 | 86 | 348029.616000 |
| 37 | 149733.672000 | 87 | 352076.472000 |
| 38 | 153780.528000 | 88 | 356123.328000 |
| 39 | 157827.384000 | 89 | 360170.184000 |
| 40 | 161874.240000 | 90 | 364217.040000 |
| 41 | 165921.096000 | 91 | 368263.896000 |
| 42 | 169967.952000 | 92 | 372310.752000 |
| 43 | 174014.808000 | 93 | 376357.608000 |
| 44 | 178061.664000 | 94 | 380404.464000 |
| 45 | 182108.520000 | 95 | 384451.320000 |
| 46 | 186155.376000 | 96 | 388498.176000 |
| 47 | 190202.232000 | 97 | 392545.032000 |
| 48 | 194249.088000 | 98 | 396591.888000 |
| 49 | 198295.944000 | 99 | 400638.744000 |
| 50 | 202342.800000 | 100 | 404685.600000 |

## AREA
### Square Meters to Acres

| m² | a | m² | a |
|---|---|---|---|
| 1 | .000247 | 51 | .012597 |
| 2 | .000494 | 52 | .012844 |
| 3 | .000741 | 53 | .013091 |
| 4 | .000988 | 54 | .013338 |
| 5 | .001235 | 55 | .013585 |
| 6 | .001482 | 56 | .013832 |
| 7 | .001729 | 57 | .014079 |
| 8 | .001976 | 58 | .014326 |
| 9 | .002223 | 59 | .014573 |
| 10 | .002470 | 60 | .014820 |
| 11 | .002717 | 61 | .015067 |
| 12 | .002964 | 62 | .015314 |
| 13 | .003211 | 63 | .015561 |
| 14 | .003458 | 64 | .015808 |
| 15 | .003705 | 65 | .016055 |
| 16 | .003952 | 66 | .016302 |
| 17 | .004199 | 67 | .016549 |
| 18 | .004446 | 68 | .016796 |
| 19 | .004693 | 69 | .017043 |
| 20 | .004940 | 70 | .017290 |
| 21 | .005187 | 71 | .017537 |
| 22 | .005434 | 72 | .017784 |
| 23 | .005681 | 73 | .018031 |
| 24 | .005928 | 74 | .018278 |
| 25 | .006175 | 75 | .018525 |
| 26 | .006422 | 76 | .018772 |
| 27 | .006669 | 77 | .019019 |
| 28 | .006916 | 78 | .019266 |
| 29 | .007163 | 79 | .019513 |
| 30 | .007410 | 80 | .019760 |
| 31 | .007657 | 81 | .020007 |
| 32 | .007904 | 82 | .020254 |
| 33 | .008151 | 83 | .020501 |
| 34 | .008398 | 84 | .020748 |
| 35 | .008645 | 85 | .020995 |
| 36 | .008892 | 86 | .021242 |
| 37 | .009139 | 87 | .021489 |
| 38 | .009386 | 88 | .021736 |
| 39 | .009633 | 89 | .021983 |
| 40 | .009880 | 90 | .022230 |
| 41 | .010127 | 91 | .022477 |
| 42 | .010374 | 92 | .022724 |
| 43 | .010621 | 93 | .022971 |
| 44 | .010868 | 94 | .023218 |
| 45 | .011115 | 95 | .023465 |
| 46 | .011362 | 96 | .023712 |
| 47 | .011609 | 97 | .023959 |
| 48 | .011856 | 98 | .024206 |
| 49 | .012103 | 99 | .024453 |
| 50 | .012350 | 100 | .024700 |

## AREA
### Acres to Hectares

| a | ha | a | ha |
|---|---|---|---|
| 1 | .404686 | 51 | 20.638986 |
| 2 | .809372 | 52 | 21.043672 |
| 3 | 1.214058 | 53 | 21.448358 |
| 4 | 1.618744 | 54 | 21.853044 |
| 5 | 2.023430 | 55 | 22.257730 |
| 6 | 2.428116 | 56 | 22.662416 |
| 7 | 2.832802 | 57 | 23.067102 |
| 8 | 3.237488 | 58 | 23.471788 |
| 9 | 3.642174 | 59 | 23.876474 |
| 10 | 4.046860 | 60 | 24.281160 |
| 11 | 4.451546 | 61 | 24.685846 |
| 12 | 4.856232 | 62 | 25.090532 |
| 13 | 5.260918 | 63 | 25.495218 |
| 14 | 5.665604 | 64 | 25.899904 |
| 15 | 6.070290 | 65 | 26.304590 |
| 16 | 6.474976 | 66 | 26.709276 |
| 17 | 6.879662 | 67 | 27.113962 |
| 18 | 7.284348 | 68 | 27.518648 |
| 19 | 7.689034 | 69 | 27.923334 |
| 20 | 8.093720 | 70 | 28.328020 |
| 21 | 8.498406 | 71 | 28.732706 |
| 22 | 8.903092 | 72 | 29.137392 |
| 23 | 9.307778 | 73 | 29.542078 |
| 24 | 9.712464 | 74 | 29.946764 |
| 25 | 10.117150 | 75 | 30.351450 |
| 26 | 10.521836 | 76 | 30.756136 |
| 27 | 10.926522 | 77 | 31.160822 |
| 28 | 11.331208 | 78 | 31.565508 |
| 29 | 11.735894 | 79 | 31.970194 |
| 30 | 12.140580 | 80 | 32.374880 |
| 31 | 12.545266 | 81 | 32.779566 |
| 32 | 12.949952 | 82 | 33.184252 |
| 33 | 13.354638 | 83 | 33.588938 |
| 34 | 13.759324 | 84 | 33.993624 |
| 35 | 14.164010 | 85 | 34.398310 |
| 36 | 14.568696 | 86 | 34.802996 |
| 37 | 14.973382 | 87 | 35.207682 |
| 38 | 15.378068 | 88 | 35.612368 |
| 39 | 15.782754 | 89 | 36.017054 |
| 40 | 16.187440 | 90 | 36.421740 |
| 41 | 16.592126 | 91 | 36.826426 |
| 42 | 16.996812 | 92 | 37.231112 |
| 43 | 17.401498 | 93 | 37.635798 |
| 44 | 17.806184 | 94 | 38.040484 |
| 45 | 18.210870 | 95 | 38.445170 |
| 46 | 18.615556 | 96 | 38.849856 |
| 47 | 19.020242 | 97 | 39.254542 |
| 48 | 19.424928 | 98 | 39.659228 |
| 49 | 19.829614 | 99 | 40.063914 |
| 50 | 20.234300 | 100 | 40.468600 |

## AREA
## Hectares to Acres

| ha | a | ha | a |
|---|---|---|---|
| 1 | 2.471054 | 51 | 126.023754 |
| 2 | 4.942108 | 52 | 128.494808 |
| 3 | 7.413162 | 53 | 130.965862 |
| 4 | 9.884216 | 54 | 133.436916 |
| 5 | 12.355270 | 55 | 135.907970 |
| 6 | 14.826324 | 56 | 138.379024 |
| 7 | 17.297378 | 57 | 140.850078 |
| 8 | 19.768432 | 58 | 143.321132 |
| 9 | 22.239486 | 59 | 145.792186 |
| 10 | 24.710540 | 60 | 148.263240 |
| 11 | 27.181594 | 61 | 150.734294 |
| 12 | 29.652648 | 62 | 153.205348 |
| 13 | 32.123702 | 63 | 155.676402 |
| 14 | 34.594756 | 64 | 158.147456 |
| 15 | 37.065810 | 65 | 160.618510 |
| 16 | 39.536864 | 66 | 163.089564 |
| 17 | 42.007918 | 67 | 165.560618 |
| 18 | 44.478972 | 68 | 168.031672 |
| 19 | 46.950026 | 69 | 170.502726 |
| 20 | 49.421080 | 70 | 172.973780 |
| 21 | 51.892134 | 71 | 175.444834 |
| 22 | 54.363188 | 72 | 177.915888 |
| 23 | 56.834242 | 73 | 180.386942 |
| 24 | 59.305296 | 74 | 182.857996 |
| 25 | 61.776350 | 75 | 185.329050 |
| 26 | 64.247404 | 76 | 187.800104 |
| 27 | 66.718458 | 77 | 190.271158 |
| 28 | 69.189512 | 78 | 192.742212 |
| 29 | 71.660566 | 79 | 195.213266 |
| 30 | 74.131620 | 80 | 197.684320 |
| 31 | 76.602674 | 81 | 200.155374 |
| 32 | 79.073728 | 82 | 202.626428 |
| 33 | 81.544782 | 83 | 205.097482 |
| 34 | 84.015836 | 84 | 207.568536 |
| 35 | 86.486890 | 85 | 210.039590 |
| 36 | 88.957944 | 86 | 212.510644 |
| 37 | 91.428998 | 87 | 214.981698 |
| 38 | 93.900052 | 88 | 217.452752 |
| 39 | 96.371106 | 89 | 219.923806 |
| 40 | 98.842160 | 90 | 222.394860 |
| 41 | 101.313214 | 91 | 224.865914 |
| 42 | 103.784268 | 92 | 227.336968 |
| 43 | 106.255322 | 93 | 229.808022 |
| 44 | 108.726376 | 94 | 232.279076 |
| 45 | 111.197430 | 95 | 234.750130 |
| 46 | 113.668484 | 96 | 237.221184 |
| 47 | 116.139538 | 97 | 239.692238 |
| 48 | 118.610592 | 98 | 242.163292 |
| 49 | 121.081646 | 99 | 244.634346 |
| 50 | 123.552700 | 100 | 247.105400 |

## AREA
## Square Miles to Square Kilometers

| mi² | km² | mi² | km² |
|---|---|---|---|
| 1 | 2.589988 | 51 | 132.089388 |
| 2 | 5.179976 | 52 | 134.679376 |
| 3 | 7.769964 | 53 | 137.269364 |
| 4 | 10.359952 | 54 | 139.859352 |
| 5 | 12.949940 | 55 | 142.449340 |
| 6 | 15.539928 | 56 | 145.039328 |
| 7 | 18.129916 | 57 | 147.629316 |
| 8 | 20.719904 | 58 | 150.219304 |
| 9 | 23.309892 | 59 | 152.809292 |
| 10 | 25.899880 | 60 | 155.399280 |
| 11 | 28.489868 | 61 | 157.989268 |
| 12 | 31.079856 | 62 | 160.579256 |
| 13 | 33.669844 | 63 | 163.169244 |
| 14 | 36.259832 | 64 | 165.759232 |
| 15 | 38.849820 | 65 | 168.349220 |
| 16 | 41.439808 | 66 | 170.939208 |
| 17 | 44.029796 | 67 | 173.529196 |
| 18 | 46.619784 | 68 | 176.119184 |
| 19 | 49.209772 | 69 | 178.709172 |
| 20 | 51.799760 | 70 | 181.299160 |
| 21 | 54.389748 | 71 | 183.889148 |
| 22 | 56.979736 | 72 | 186.479136 |
| 23 | 59.569724 | 73 | 189.069124 |
| 24 | 62.159712 | 74 | 191.659112 |
| 25 | 64.749700 | 75 | 194.249100 |
| 26 | 67.339688 | 76 | 196.839088 |
| 27 | 69.929676 | 77 | 199.429076 |
| 28 | 72.519664 | 78 | 202.019064 |
| 29 | 75.109652 | 79 | 204.609052 |
| 30 | 77.699640 | 80 | 207.199040 |
| 31 | 80.289628 | 81 | 209.789028 |
| 32 | 82.879616 | 82 | 212.379016 |
| 33 | 85.469604 | 83 | 214.969004 |
| 34 | 88.059592 | 84 | 217.558992 |
| 35 | 90.649580 | 85 | 220.148980 |
| 36 | 93.239568 | 86 | 222.738968 |
| 37 | 95.829556 | 87 | 225.328956 |
| 38 | 98.419544 | 88 | 227.918944 |
| 39 | 101.009532 | 89 | 230.508932 |
| 40 | 103.599520 | 90 | 233.098920 |
| 41 | 106.189508 | 91 | 235.688908 |
| 42 | 108.779496 | 92 | 238.278896 |
| 43 | 111.369484 | 93 | 240.868884 |
| 44 | 113.959472 | 94 | 243.458872 |
| 45 | 116.549460 | 95 | 246.048860 |
| 46 | 119.139448 | 96 | 248.638848 |
| 47 | 121.729436 | 97 | 251.228836 |
| 48 | 124.319424 | 98 | 253.818824 |
| 49 | 126.909412 | 99 | 256.408812 |
| 50 | 129.499400 | 100 | 258.998800 |

## AREA
### Square Kilometers to Square Miles

| km² | mi² | km² | mi² |
|---|---|---|---|
| 1 | .386102 | 51 | 19.691202 |
| 2 | .772204 | 52 | 20.077304 |
| 3 | 1.158306 | 53 | 20.463406 |
| 4 | 1.544408 | 54 | 20.849508 |
| 5 | 1.930510 | 55 | 21.235610 |
| 6 | 2.316612 | 56 | 21.621712 |
| 7 | 2.702714 | 57 | 22.007814 |
| 8 | 3.088816 | 58 | 22.393916 |
| 9 | 3.474918 | 59 | 22.780018 |
| 10 | 3.861020 | 60 | 23.166120 |
| 11 | 4.247122 | 61 | 23.552222 |
| 12 | 4.633224 | 62 | 23.938324 |
| 13 | 5.019326 | 63 | 24.324426 |
| 14 | 5.405428 | 64 | 24.710528 |
| 15 | 5.791530 | 65 | 25.096630 |
| 16 | 6.177632 | 66 | 25.482732 |
| 17 | 6.563734 | 67 | 25.868834 |
| 18 | 6.949836 | 68 | 26.254936 |
| 19 | 7.335938 | 69 | 26.641038 |
| 20 | 7.722040 | 70 | 27.027140 |
| 21 | 8.108142 | 71 | 27.413242 |
| 22 | 8.494244 | 72 | 27.799344 |
| 23 | 8.880346 | 73 | 28.185446 |
| 24 | 9.266448 | 74 | 28.571548 |
| 25 | 9.652550 | 75 | 28.957650 |
| 26 | 10.038652 | 76 | 29.343752 |
| 27 | 10.424754 | 77 | 29.729854 |
| 28 | 10.810856 | 78 | 30.115956 |
| 29 | 11.196958 | 79 | 30.502058 |
| 30 | 11.583060 | 80 | 30.888160 |
| 31 | 11.969162 | 81 | 31.274262 |
| 32 | 12.355264 | 82 | 31.660364 |
| 33 | 12.741366 | 83 | 32.046466 |
| 34 | 13.127468 | 84 | 32.432568 |
| 35 | 13.513570 | 85 | 32.818670 |
| 36 | 13.899672 | 86 | 33.204772 |
| 37 | 14.285774 | 87 | 33.590874 |
| 38 | 14.671876 | 88 | 33.976976 |
| 39 | 15.057978 | 89 | 34.363078 |
| 40 | 15.444080 | 90 | 34.749180 |
| 41 | 15.830182 | 91 | 35.135282 |
| 42 | 16.216284 | 92 | 35.521384 |
| 43 | 16.602386 | 93 | 35.907486 |
| 44 | 16.988488 | 94 | 36.293588 |
| 45 | 17.374590 | 95 | 36.679690 |
| 46 | 17.760692 | 96 | 37.065792 |
| 47 | 18.146794 | 97 | 37.451894 |
| 48 | 18.532896 | 98 | 37.837996 |
| 49 | 18.918998 | 99 | 38.224098 |
| 50 | 19.305100 | 100 | 38.610200 |

## VOLUME
### Cubic Inches to Cubic Centimeters

| in³ | cm³ | in³ | cm³ |
|---|---|---|---|
| 1 | 16.387060 | 51 | 835.740060 |
| 2 | 32.774120 | 52 | 852.127120 |
| 3 | 49.161180 | 53 | 868.514180 |
| 4 | 65.548240 | 54 | 884.901240 |
| 5 | 81.935300 | 55 | 901.288300 |
| 6 | 98.322360 | 56 | 917.675360 |
| 7 | 114.709420 | 57 | 934.062420 |
| 8 | 131.096480 | 58 | 950.449480 |
| 9 | 147.483540 | 59 | 966.836540 |
| 10 | 163.870600 | 60 | 983.223600 |
| 11 | 180.257660 | 61 | 999.610660 |
| 12 | 196.644720 | 62 | 1015.997720 |
| 13 | 213.031780 | 63 | 1032.384780 |
| 14 | 229.418840 | 64 | 1048.771840 |
| 15 | 245.805900 | 65 | 1065.158900 |
| 16 | 262.192960 | 66 | 1081.545960 |
| 17 | 278.580020 | 67 | 1097.933020 |
| 18 | 294.967080 | 68 | 1114.320080 |
| 19 | 311.354140 | 69 | 1130.707140 |
| 20 | 327.741200 | 70 | 1147.094200 |
| 21 | 344.128260 | 71 | 1163.481260 |
| 22 | 360.515320 | 72 | 1179.868320 |
| 23 | 376.902380 | 73 | 1196.255380 |
| 24 | 393.289440 | 74 | 1212.642440 |
| 25 | 409.676500 | 75 | 1229.029500 |
| 26 | 426.063560 | 76 | 1245.416560 |
| 27 | 442.450620 | 77 | 1261.803620 |
| 28 | 458.837680 | 78 | 1278.190680 |
| 29 | 475.224740 | 79 | 1294.577740 |
| 30 | 491.611800 | 80 | 1310.964800 |
| 31 | 507.998860 | 81 | 1327.351860 |
| 32 | 524.385920 | 82 | 1343.738920 |
| 33 | 540.772980 | 83 | 1360.125980 |
| 34 | 557.160040 | 84 | 1376.513040 |
| 35 | 573.547100 | 85 | 1392.900100 |
| 36 | 589.934160 | 86 | 1409.287160 |
| 37 | 606.321220 | 87 | 1425.674220 |
| 38 | 622.708280 | 88 | 1442.061280 |
| 39 | 639.095340 | 89 | 1458.448340 |
| 40 | 655.482400 | 90 | 1474.835400 |
| 41 | 671.869460 | 91 | 1491.222460 |
| 42 | 688.256520 | 92 | 1507.609520 |
| 43 | 704.643580 | 93 | 1523.996580 |
| 44 | 721.030640 | 94 | 1540.383640 |
| 45 | 737.417700 | 95 | 1556.770700 |
| 46 | 753.804760 | 96 | 1573.157760 |
| 47 | 770.191820 | 97 | 1589.544820 |
| 48 | 786.578880 | 98 | 1605.931880 |
| 49 | 802.965940 | 99 | 1622.318940 |
| 50 | 819.353000 | 100 | 1638.706000 |

## VOLUME
### Cubic Centimeters to Cubic Inches

| cm³ | in³ | cm³ | in³ |
|---|---|---|---|
| 1 | .061023 | 51 | 3.112173 |
| 2 | .122046 | 52 | 3.173196 |
| 3 | .183069 | 53 | 3.234219 |
| 4 | .244092 | 54 | 3.295242 |
| 5 | .305115 | 55 | 3.356265 |
| 6 | .366138 | 56 | 3.417288 |
| 7 | .427161 | 57 | 3.478311 |
| 8 | .488184 | 58 | 3.539334 |
| 9 | .549207 | 59 | 3.600357 |
| 10 | .610230 | 60 | 3.661380 |
| 11 | .671253 | 61 | 3.722403 |
| 12 | .732276 | 62 | 3.783426 |
| 13 | .793299 | 63 | 3.844449 |
| 14 | .854322 | 64 | 3.905472 |
| 15 | .915345 | 65 | 3.966495 |
| 16 | .976368 | 66 | 4.027518 |
| 17 | 1.037391 | 67 | 4.088541 |
| 18 | 1.098414 | 68 | 4.149564 |
| 19 | 1.159437 | 69 | 4.210587 |
| 20 | 1.220460 | 70 | 4.271610 |
| 21 | 1.281483 | 71 | 4.332633 |
| 22 | 1.342506 | 72 | 4.393656 |
| 23 | 1.403529 | 73 | 4.454679 |
| 24 | 1.464552 | 74 | 4.515702 |
| 25 | 1.525575 | 75 | 4.576725 |
| 26 | 1.586598 | 76 | 4.637748 |
| 27 | 1.647621 | 77 | 4.698771 |
| 28 | 1.708644 | 78 | 4.759794 |
| 29 | 1.769667 | 79 | 4.820817 |
| 30 | 1.830690 | 80 | 4.881840 |
| 31 | 1.891713 | 81 | 4.942863 |
| 32 | 1.952736 | 82 | 5.003886 |
| 33 | 2.013759 | 83 | 5.064909 |
| 34 | 2.074782 | 84 | 5.125932 |
| 35 | 2.135805 | 85 | 5.186955 |
| 36 | 2.196828 | 86 | 5.247978 |
| 37 | 2.257851 | 87 | 5.309001 |
| 38 | 2.318874 | 88 | 5.370024 |
| 39 | 2.379897 | 89 | 5.431047 |
| 40 | 2.440920 | 90 | 5.492070 |
| 41 | 2.501943 | 91 | 5.553093 |
| 42 | 2.562966 | 92 | 5.614116 |
| 43 | 2.623989 | 93 | 5.675139 |
| 44 | 2.685012 | 94 | 5.736162 |
| 45 | 2.746035 | 95 | 5.797185 |
| 46 | 2.807058 | 96 | 5.858208 |
| 47 | 2.868081 | 97 | 5.919231 |
| 48 | 2.929104 | 98 | 5.980254 |
| 49 | 2.990127 | 99 | 6.041277 |
| 50 | 3.051150 | 100 | 6.102300 |

## VOLUME
### Cubic Feet to Cubic Meters

| ft³ | m³ | ft³ | m³ |
|---|---|---|---|
| 1 | .028316 | 51 | 1.444116 |
| 2 | .056632 | 52 | 1.472432 |
| 3 | .084948 | 53 | 1.500748 |
| 4 | .113264 | 54 | 1.529064 |
| 5 | .141580 | 55 | 1.557380 |
| 6 | .169896 | 56 | 1.585696 |
| 7 | .198212 | 57 | 1.614012 |
| 8 | .226528 | 58 | 1.642328 |
| 9 | .254844 | 59 | 1.670644 |
| 10 | .283160 | 60 | 1.698960 |
| 11 | .311476 | 61 | 1.727276 |
| 12 | .339792 | 62 | 1.755592 |
| 13 | .368108 | 63 | 1.783908 |
| 14 | .396424 | 64 | 1.812224 |
| 15 | .424740 | 65 | 1.840540 |
| 16 | .453056 | 66 | 1.868856 |
| 17 | .481372 | 67 | 1.897172 |
| 18 | .509688 | 68 | 1.925488 |
| 19 | .538004 | 69 | 1.953804 |
| 20 | .566320 | 70 | 1.982120 |
| 21 | .594636 | 71 | 2.010436 |
| 22 | .622952 | 72 | 2.038752 |
| 23 | .651268 | 73 | 2.067068 |
| 24 | .679584 | 74 | 2.095384 |
| 25 | .707900 | 75 | 2.123700 |
| 26 | .736216 | 76 | 2.152016 |
| 27 | .764532 | 77 | 2.180332 |
| 28 | .792848 | 78 | 2.208648 |
| 29 | .821164 | 79 | 2.236964 |
| 30 | .849480 | 80 | 2.265280 |
| 31 | .877796 | 81 | 2.293596 |
| 32 | .906112 | 82 | 2.321912 |
| 33 | .934428 | 83 | 2.350228 |
| 34 | .962744 | 84 | 2.378544 |
| 35 | .991060 | 85 | 2.406860 |
| 36 | 1.019376 | 86 | 2.435176 |
| 37 | 1.047692 | 87 | 2.463492 |
| 38 | 1.076008 | 88 | 2.491808 |
| 39 | 1.104324 | 89 | 2.520124 |
| 40 | 1.132640 | 90 | 2.548440 |
| 41 | 1.160956 | 91 | 2.576756 |
| 42 | 1.189272 | 92 | 2.605072 |
| 43 | 1.217588 | 93 | 2.633388 |
| 44 | 1.245904 | 94 | 2.661704 |
| 45 | 1.274220 | 95 | 2.690020 |
| 46 | 1.302536 | 96 | 2.718336 |
| 47 | 1.330852 | 97 | 2.746652 |
| 48 | 1.359168 | 98 | 2.774968 |
| 49 | 1.387484 | 99 | 2.803284 |
| 50 | 1.415800 | 100 | 2.831600 |

## VOLUME
### Cubic Meters to Cubic Feet

| m³ | ft³ | m³ | ft³ |
|---|---|---|---|
| 1 | 35.314660 | 51 | 1801.047660 |
| 2 | 70.629320 | 52 | 1836.362320 |
| 3 | 105.943980 | 53 | 1871.676980 |
| 4 | 141.258640 | 54 | 1906.991640 |
| 5 | 176.573300 | 55 | 1942.306300 |
| 6 | 211.887960 | 56 | 1977.620960 |
| 7 | 247.202620 | 57 | 2012.935620 |
| 8 | 282.517280 | 58 | 2048.250280 |
| 9 | 317.831940 | 59 | 2083.564940 |
| 10 | 353.146600 | 60 | 2118.879600 |
| 11 | 388.461260 | 61 | 2154.194260 |
| 12 | 423.775920 | 62 | 2189.508920 |
| 13 | 459.090580 | 63 | 2224.823580 |
| 14 | 494.405240 | 64 | 2260.138240 |
| 15 | 529.719900 | 65 | 2295.452900 |
| 16 | 565.034560 | 66 | 2330.767560 |
| 17 | 600.349220 | 67 | 2366.082220 |
| 18 | 635.663880 | 68 | 2401.396880 |
| 19 | 670.978540 | 69 | 2436.711540 |
| 20 | 706.293200 | 70 | 2472.026200 |
| 21 | 741.607860 | 71 | 2507.340860 |
| 22 | 776.922520 | 72 | 2542.655520 |
| 23 | 812.237180 | 73 | 2577.970180 |
| 24 | 847.551840 | 74 | 2613.284840 |
| 25 | 882.866500 | 75 | 2648.599500 |
| 26 | 918.181160 | 76 | 2683.914160 |
| 27 | 953.495820 | 77 | 2719.228820 |
| 28 | 988.810480 | 78 | 2754.543480 |
| 29 | 1024.125140 | 79 | 2789.858140 |
| 30 | 1059.439800 | 80 | 2825.172800 |
| 31 | 1094.754460 | 81 | 2860.487460 |
| 32 | 1130.069120 | 82 | 2895.802120 |
| 33 | 1165.383780 | 83 | 2931.116780 |
| 34 | 1200.698440 | 84 | 2966.431440 |
| 35 | 1236.013100 | 85 | 3001.746100 |
| 36 | 1271.327760 | 86 | 3037.060760 |
| 37 | 1306.642420 | 87 | 3072.375420 |
| 38 | 1341.957080 | 88 | 3107.690080 |
| 39 | 1377.271740 | 89 | 3143.004740 |
| 40 | 1412.586400 | 90 | 3178.319400 |
| 41 | 1447.901060 | 91 | 3213.634060 |
| 42 | 1483.215720 | 92 | 3248.948720 |
| 43 | 1518.530380 | 93 | 3284.263380 |
| 44 | 1553.845040 | 94 | 3319.578040 |
| 45 | 1589.159700 | 95 | 3354.892700 |
| 46 | 1624.474360 | 96 | 3390.207360 |
| 47 | 1659.789020 | 97 | 3425.522020 |
| 48 | 1695.103680 | 98 | 3460.836680 |
| 49 | 1730.418340 | 99 | 3496.151340 |
| 50 | 1765.733000 | 100 | 3531.466000 |

## VOLUME
### Cubic Yards to Cubic Meters

| yd³ | m³ | yd³ | m³ |
|---|---|---|---|
| 1 | .764554 | 51 | 38.992254 |
| 2 | 1.529108 | 52 | 39.756808 |
| 3 | 2.293662 | 53 | 40.521362 |
| 4 | 3.058216 | 54 | 41.285916 |
| 5 | 3.822770 | 55 | 42.050470 |
| 6 | 4.587324 | 56 | 42.815024 |
| 7 | 5.351878 | 57 | 43.579578 |
| 8 | 6.116432 | 58 | 44.344132 |
| 9 | 6.880986 | 59 | 45.108686 |
| 10 | 7.645540 | 60 | 45.873240 |
| 11 | 8.410094 | 61 | 46.637794 |
| 12 | 9.174648 | 62 | 47.402348 |
| 13 | 9.939202 | 63 | 48.166902 |
| 14 | 10.703756 | 64 | 48.931456 |
| 15 | 11.468310 | 65 | 49.696010 |
| 16 | 12.232864 | 66 | 50.460564 |
| 17 | 12.997418 | 67 | 51.225118 |
| 18 | 13.761972 | 68 | 51.989672 |
| 19 | 14.526526 | 69 | 52.754226 |
| 20 | 15.291080 | 70 | 53.518780 |
| 21 | 16.055634 | 71 | 54.283334 |
| 22 | 16.820188 | 72 | 55.047888 |
| 23 | 17.584742 | 73 | 55.812442 |
| 24 | 18.349296 | 74 | 56.576996 |
| 25 | 19.113850 | 75 | 57.341550 |
| 26 | 19.878404 | 76 | 58.106104 |
| 27 | 20.642958 | 77 | 58.870658 |
| 28 | 21.407512 | 78 | 59.635212 |
| 29 | 22.172066 | 79 | 60.399766 |
| 30 | 22.936620 | 80 | 61.164320 |
| 31 | 23.701174 | 81 | 61.928874 |
| 32 | 24.465728 | 82 | 62.693428 |
| 33 | 25.230282 | 83 | 63.457982 |
| 34 | 25.994836 | 84 | 64.222536 |
| 35 | 26.759390 | 85 | 64.987090 |
| 36 | 27.523944 | 86 | 65.751644 |
| 37 | 28.288498 | 87 | 66.516198 |
| 38 | 29.053052 | 88 | 67.280752 |
| 39 | 29.817606 | 89 | 68.045306 |
| 40 | 30.582160 | 90 | 68.809860 |
| 41 | 31.346714 | 91 | 69.574414 |
| 42 | 32.111268 | 92 | 70.338968 |
| 43 | 32.875822 | 93 | 71.103522 |
| 44 | 33.640376 | 94 | 71.868076 |
| 45 | 34.404930 | 95 | 72.632630 |
| 46 | 35.169484 | 96 | 73.397184 |
| 47 | 35.934038 | 97 | 74.161738 |
| 48 | 36.698592 | 98 | 74.926292 |
| 49 | 37.463146 | 99 | 75.690846 |
| 50 | 38.227700 | 100 | 76.455400 |

## VOLUME
### Cubic Meters to Cubic Yards

| m³ | yd³ | m³ | yd³ |
|---|---|---|---|
| 1 | 1.307950 | 51 | 66.705450 |
| 2 | 2.615900 | 52 | 68.013400 |
| 3 | 3.923850 | 53 | 69.321350 |
| 4 | 5.231800 | 54 | 70.629300 |
| 5 | 6.539750 | 55 | 71.937250 |
| 6 | 7.847700 | 56 | 73.245200 |
| 7 | 9.155650 | 57 | 74.553150 |
| 8 | 10.463600 | 58 | 75.861100 |
| 9 | 11.771550 | 59 | 77.169050 |
| 10 | 13.079500 | 60 | 78.477000 |
| 11 | 14.387450 | 61 | 79.784950 |
| 12 | 15.695400 | 62 | 81.092900 |
| 13 | 17.003350 | 63 | 82.400850 |
| 14 | 18.311300 | 64 | 83.708800 |
| 15 | 19.619250 | 65 | 85.016750 |
| 16 | 20.927200 | 66 | 86.324700 |
| 17 | 22.235150 | 67 | 87.632650 |
| 18 | 23.543100 | 68 | 88.940600 |
| 19 | 24.851050 | 69 | 90.248550 |
| 20 | 26.159000 | 70 | 91.556500 |
| 21 | 27.466950 | 71 | 92.864450 |
| 22 | 28.774900 | 72 | 94.172400 |
| 23 | 30.082850 | 73 | 95.480350 |
| 24 | 31.390800 | 74 | 96.788300 |
| 25 | 32.698750 | 75 | 98.096250 |
| 26 | 34.006700 | 76 | 99.404200 |
| 27 | 35.314650 | 77 | 100.712150 |
| 28 | 36.622600 | 78 | 102.020100 |
| 29 | 37.930550 | 79 | 103.328050 |
| 30 | 39.238500 | 80 | 104.636000 |
| 31 | 40.546450 | 81 | 105.943950 |
| 32 | 41.854400 | 82 | 107.251900 |
| 33 | 43.162350 | 83 | 108.559850 |
| 34 | 44.470300 | 84 | 109.867800 |
| 35 | 45.778250 | 85 | 111.175750 |
| 36 | 47.086200 | 86 | 112.483700 |
| 37 | 48.394150 | 87 | 113.791650 |
| 38 | 49.702100 | 88 | 115.099600 |
| 39 | 51.010050 | 89 | 116.407550 |
| 40 | 52.318000 | 90 | 117.715500 |
| 41 | 53.625950 | 91 | 119.023450 |
| 42 | 54.933900 | 92 | 120.331400 |
| 43 | 56.241850 | 93 | 121.639350 |
| 44 | 57.549800 | 94 | 122.947300 |
| 45 | 58.857750 | 95 | 124.255250 |
| 46 | 60.165700 | 96 | 125.563200 |
| 47 | 61.473650 | 97 | 126.871150 |
| 48 | 62.781600 | 98 | 128.179100 |
| 49 | 64.089550 | 99 | 129.487050 |
| 50 | 65.397500 | 100 | 130.795000 |

## VOLUME
### Ounces (fluid) to Milliliters

| oz | ml | oz | ml |
|---|---|---|---|
| 1 | 29.573529 | 51 | 1508.249979 |
| 2 | 59.147058 | 52 | 1537.823508 |
| 3 | 88.720587 | 53 | 1567.397037 |
| 4 | 118.294116 | 54 | 1596.970566 |
| 5 | 147.867645 | 55 | 1626.544095 |
| 6 | 177.441174 | 56 | 1656.117624 |
| 7 | 207.014703 | 57 | 1685.691153 |
| 8 | 236.588232 | 58 | 1715.264682 |
| 9 | 266.161761 | 59 | 1744.838211 |
| 10 | 295.735290 | 60 | 1774.411740 |
| 11 | 325.308819 | 61 | 1803.985269 |
| 12 | 354.882348 | 62 | 1833.558798 |
| 13 | 384.455877 | 63 | 1863.132327 |
| 14 | 414.029406 | 64 | 1892.705856 |
| 15 | 443.602935 | 65 | 1922.279385 |
| 16 | 473.176464 | 66 | 1951.852914 |
| 17 | 502.749993 | 67 | 1981.426443 |
| 18 | 532.323522 | 68 | 2010.999972 |
| 19 | 561.897051 | 69 | 2040.573501 |
| 20 | 591.470580 | 70 | 2070.147030 |
| 21 | 621.044109 | 71 | 2099.720559 |
| 22 | 650.617638 | 72 | 2129.294088 |
| 23 | 680.191167 | 73 | 2158.867617 |
| 24 | 709.764696 | 74 | 2188.441146 |
| 25 | 739.338225 | 75 | 2218.014675 |
| 26 | 768.911754 | 76 | 2247.588204 |
| 27 | 798.485283 | 77 | 2277.161733 |
| 28 | 828.058812 | 78 | 2306.735262 |
| 29 | 857.632341 | 79 | 2336.308791 |
| 30 | 887.205870 | 80 | 2365.882320 |
| 31 | 916.779399 | 81 | 2395.455849 |
| 32 | 946.352928 | 82 | 2425.029378 |
| 33 | 975.926457 | 83 | 2454.602907 |
| 34 | 1005.499986 | 84 | 2484.176436 |
| 35 | 1035.073515 | 85 | 2513.749965 |
| 36 | 1064.647044 | 86 | 2543.323494 |
| 37 | 1094.220573 | 87 | 2572.897023 |
| 38 | 1123.794102 | 88 | 2602.470552 |
| 39 | 1153.367631 | 89 | 2632.044081 |
| 40 | 1182.941160 | 90 | 2661.617610 |
| 41 | 1212.514689 | 91 | 2691.191139 |
| 42 | 1242.088218 | 92 | 2720.764668 |
| 43 | 1271.661747 | 93 | 2750.338197 |
| 44 | 1301.235276 | 94 | 2779.911726 |
| 45 | 1330.808805 | 95 | 2809.485255 |
| 46 | 1360.382334 | 96 | 2839.058784 |
| 47 | 1389.955863 | 97 | 2868.632313 |
| 48 | 1419.529392 | 98 | 2898.205842 |
| 49 | 1449.102921 | 99 | 2927.779371 |
| 50 | 1478.676450 | 100 | 2957.352900 |

## VOLUME
### Milliliters to Ounces (fluid)

| ml | oz | ml | oz |
|---|---|---|---|
| 1 | .033814 | 51 | 1.724514 |
| 2 | .067628 | 52 | 1.758328 |
| 3 | .101442 | 53 | 1.792142 |
| 4 | .135256 | 54 | 1.825956 |
| 5 | .169070 | 55 | 1.859770 |
| 6 | .202884 | 56 | 1.893584 |
| 7 | .236698 | 57 | 1.927398 |
| 8 | .270512 | 58 | 1.961212 |
| 9 | .304326 | 59 | 1.995026 |
| 10 | .338140 | 60 | 2.028840 |
| 11 | .371954 | 61 | 2.062654 |
| 12 | .405768 | 62 | 2.096468 |
| 13 | .439582 | 63 | 2.130282 |
| 14 | .473396 | 64 | 2.164096 |
| 15 | .507210 | 65 | 2.197910 |
| 16 | .541024 | 66 | 2.231724 |
| 17 | .574838 | 67 | 2.265538 |
| 18 | .608652 | 68 | 2.299352 |
| 19 | .642466 | 69 | 2.333166 |
| 20 | .676280 | 70 | 2.366980 |
| 21 | .710094 | 71 | 2.400794 |
| 22 | .743908 | 72 | 2.434608 |
| 23 | .777722 | 73 | 2.468422 |
| 24 | .811536 | 74 | 2.502236 |
| 25 | .845350 | 75 | 2.536050 |
| 26 | .879164 | 76 | 2.569864 |
| 27 | .912978 | 77 | 2.603678 |
| 28 | .946792 | 78 | 2.637492 |
| 29 | .980606 | 79 | 2.671306 |
| 30 | 1.014420 | 80 | 2.705120 |
| 31 | 1.048234 | 81 | 2.738934 |
| 32 | 1.082048 | 82 | 2.772748 |
| 33 | 1.115862 | 83 | 2.806562 |
| 34 | 1.149676 | 84 | 2.840376 |
| 35 | 1.183490 | 85 | 2.874190 |
| 36 | 1.217304 | 86 | 2.908004 |
| 37 | 1.251118 | 87 | 2.941818 |
| 38 | 1.284932 | 88 | 2.975632 |
| 39 | 1.318746 | 89 | 3.009446 |
| 40 | 1.352560 | 90 | 3.043260 |
| 41 | 1.386374 | 91 | 3.077074 |
| 42 | 1.420188 | 92 | 3.110888 |
| 43 | 1.454002 | 93 | 3.144702 |
| 44 | 1.487816 | 94 | 3.178516 |
| 45 | 1.521630 | 95 | 3.212330 |
| 46 | 1.555444 | 96 | 3.246144 |
| 47 | 1.589258 | 97 | 3.279958 |
| 48 | 1.623072 | 98 | 3.313772 |
| 49 | 1.656886 | 99 | 3.347586 |
| 50 | 1.690700 | 100 | 3.381400 |

## VOLUME
### Pints (dry) to Cubic Centimeters

| pt | cm³ | pt | cm³ |
|---|---|---|---|
| 1 | 550.610500 | 51 | 28081.135500 |
| 2 | 1101.221000 | 52 | 28631.746000 |
| 3 | 1651.831500 | 53 | 29182.356500 |
| 4 | 2202.442000 | 54 | 29732.967000 |
| 5 | 2753.052500 | 55 | 30283.577500 |
| 6 | 3303.663000 | 56 | 30834.188000 |
| 7 | 3854.273500 | 57 | 31384.798500 |
| 8 | 4404.884000 | 58 | 31935.409000 |
| 9 | 4955.494500 | 59 | 32486.019500 |
| 10 | 5506.105000 | 60 | 33036.630000 |
| 11 | 6056.715500 | 61 | 33587.240500 |
| 12 | 6607.326000 | 62 | 34137.851000 |
| 13 | 7157.936500 | 63 | 34688.461500 |
| 14 | 7708.547000 | 64 | 35239.072000 |
| 15 | 8259.157500 | 65 | 35789.682500 |
| 16 | 8809.768000 | 66 | 36340.293000 |
| 17 | 9360.378500 | 67 | 36890.903500 |
| 18 | 9910.989000 | 68 | 37441.514000 |
| 19 | 10461.599500 | 69 | 37992.124500 |
| 20 | 11012.210000 | 70 | 38542.735000 |
| 21 | 11562.820500 | 71 | 39093.345500 |
| 22 | 12113.431000 | 72 | 39643.956000 |
| 23 | 12664.041500 | 73 | 40194.566500 |
| 24 | 13214.652000 | 74 | 40745.177000 |
| 25 | 13765.262500 | 75 | 41295.787500 |
| 26 | 14315.873000 | 76 | 41846.398000 |
| 27 | 14866.483500 | 77 | 42397.008500 |
| 28 | 15417.094000 | 78 | 42947.619000 |
| 29 | 15967.704500 | 79 | 43498.229500 |
| 30 | 16518.315000 | 80 | 44048.840000 |
| 31 | 17068.925500 | 81 | 44599.450500 |
| 32 | 17619.536000 | 82 | 45150.061000 |
| 33 | 18170.146500 | 83 | 45700.671500 |
| 34 | 18720.757000 | 84 | 46251.282000 |
| 35 | 19271.367500 | 85 | 46801.892500 |
| 36 | 19821.978000 | 86 | 47352.503000 |
| 37 | 20372.588500 | 87 | 47903.113500 |
| 38 | 20923.199000 | 88 | 48453.724000 |
| 39 | 21473.809500 | 89 | 49004.334500 |
| 40 | 22024.420000 | 90 | 49554.945000 |
| 41 | 22575.030500 | 91 | 50105.555500 |
| 42 | 23125.641000 | 92 | 50656.166000 |
| 43 | 23676.251500 | 93 | 51206.776500 |
| 44 | 24226.862000 | 94 | 51757.387000 |
| 45 | 24777.472500 | 95 | 52307.997500 |
| 46 | 25328.083000 | 96 | 52858.608000 |
| 47 | 25878.693500 | 97 | 53409.218500 |
| 48 | 26429.304000 | 98 | 53959.829000 |
| 49 | 26979.914500 | 99 | 54510.439500 |
| 50 | 27530.525000 | 100 | 55061.050000 |

## VOLUME
### Cubic Centimeters to Pints (dry)

| cm³ | pt |
|---|---|
| 1 | .001816 |
| 2 | .003632 |
| 3 | .005448 |
| 4 | .007264 |
| 5 | .009080 |
| 6 | .010896 |
| 7 | .012712 |
| 8 | .014528 |
| 9 | .016344 |
| 10 | .018160 |
| 11 | .019976 |
| 12 | .021792 |
| 13 | .023608 |
| 14 | .025424 |
| 15 | .027240 |
| 16 | .029056 |
| 17 | .030872 |
| 18 | .032688 |
| 19 | .034504 |
| 20 | .036320 |
| 21 | .038136 |
| 22 | .039952 |
| 23 | .041768 |
| 24 | .043584 |
| 25 | .045400 |
| 26 | .047216 |
| 27 | .049032 |
| 28 | .050848 |
| 29 | .052664 |
| 30 | .054480 |
| 31 | .056296 |
| 32 | .058112 |
| 33 | .059928 |
| 34 | .061744 |
| 35 | .063560 |
| 36 | .065376 |
| 37 | .067192 |
| 38 | .069008 |
| 39 | .070824 |
| 40 | .072640 |
| 41 | .074456 |
| 42 | .076272 |
| 43 | .078088 |
| 44 | .079904 |
| 45 | .081720 |
| 46 | .083536 |
| 47 | .085352 |
| 48 | .087168 |
| 49 | .088984 |
| 50 | .090800 |
| 51 | .092616 |
| 52 | .094432 |
| 53 | .096248 |
| 54 | .098064 |
| 55 | .099880 |
| 56 | .101696 |
| 57 | .103512 |
| 58 | .105328 |
| 59 | .107144 |
| 60 | .108960 |
| 61 | .110776 |
| 62 | .112592 |
| 63 | .114408 |
| 64 | .116224 |
| 65 | .118040 |
| 66 | .119856 |
| 67 | .121672 |
| 68 | .123488 |
| 69 | .125304 |
| 70 | .127120 |
| 71 | .128936 |
| 72 | .130752 |
| 73 | .132568 |
| 74 | .134384 |
| 75 | .136200 |
| 76 | .138016 |
| 77 | .139832 |
| 78 | .141648 |
| 79 | .143464 |
| 80 | .145280 |
| 81 | .147096 |
| 82 | .148912 |
| 83 | .150728 |
| 84 | .152544 |
| 85 | .154360 |
| 86 | .156176 |
| 87 | .157992 |
| 88 | .159808 |
| 89 | .161624 |
| 90 | .163440 |
| 91 | .165256 |
| 92 | .167072 |
| 93 | .168888 |
| 94 | .170704 |
| 95 | .172520 |
| 96 | .174336 |
| 97 | .176152 |
| 98 | .177968 |
| 99 | .179784 |
| 100 | .181600 |

## VOLUME
### Pints (liquid) to Liters

| pt | L | pt | L |
|---|---|---|---|
| 1 | .473176 | 51 | 24.131976 |
| 2 | .946352 | 52 | 24.605152 |
| 3 | 1.419528 | 53 | 25.078328 |
| 4 | 1.892704 | 54 | 25.551504 |
| 5 | 2.365880 | 55 | 26.024680 |
| 6 | 2.839056 | 56 | 26.497856 |
| 7 | 3.312232 | 57 | 26.971032 |
| 8 | 3.785408 | 58 | 27.444208 |
| 9 | 4.258584 | 59 | 27.917384 |
| 10 | 4.731760 | 60 | 28.390560 |
| 11 | 5.204936 | 61 | 28.863736 |
| 12 | 5.678112 | 62 | 29.336912 |
| 13 | 6.151288 | 63 | 29.810088 |
| 14 | 6.624464 | 64 | 30.283264 |
| 15 | 7.097640 | 65 | 30.756440 |
| 16 | 7.570816 | 66 | 31.229616 |
| 17 | 8.043992 | 67 | 31.702792 |
| 18 | 8.517168 | 68 | 32.175968 |
| 19 | 8.990344 | 69 | 32.649144 |
| 20 | 9.463520 | 70 | 33.122320 |
| 21 | 9.936696 | 71 | 33.595496 |
| 22 | 10.409872 | 72 | 34.068672 |
| 23 | 10.883048 | 73 | 34.541848 |
| 24 | 11.356224 | 74 | 35.015024 |
| 25 | 11.829400 | 75 | 35.488200 |
| 26 | 12.302576 | 76 | 35.961376 |
| 27 | 12.775752 | 77 | 36.434552 |
| 28 | 13.248928 | 78 | 36.907728 |
| 29 | 13.722104 | 79 | 37.380904 |
| 30 | 14.195280 | 80 | 37.854080 |
| 31 | 14.668456 | 81 | 38.327256 |
| 32 | 15.141632 | 82 | 38.800432 |
| 33 | 15.614808 | 83 | 39.273608 |
| 34 | 16.087984 | 84 | 39.746784 |
| 35 | 16.561160 | 85 | 40.219960 |
| 36 | 17.034336 | 86 | 40.693136 |
| 37 | 17.507512 | 87 | 41.166312 |
| 38 | 17.980688 | 88 | 41.639488 |
| 39 | 18.453864 | 89 | 42.112664 |
| 40 | 18.927040 | 90 | 42.585840 |
| 41 | 19.400216 | 91 | 43.059016 |
| 42 | 19.873392 | 92 | 43.532192 |
| 43 | 20.346568 | 93 | 44.005368 |
| 44 | 20.819744 | 94 | 44.478544 |
| 45 | 21.292920 | 95 | 44.951720 |
| 46 | 21.766096 | 96 | 45.424896 |
| 47 | 22.239272 | 97 | 45.898072 |
| 48 | 22.712448 | 98 | 46.371248 |
| 49 | 23.185624 | 99 | 46.844424 |
| 50 | 23.658800 | 100 | 47.317600 |

# VOLUME
## Liters to Pints (liquid)

| L | pt | L | pt |
|---|---|---|---|
| 1 | 2.113376 | 51 | 107.782176 |
| 2 | 4.226752 | 52 | 109.895552 |
| 3 | 6.340128 | 53 | 112.008928 |
| 4 | 8.453504 | 54 | 114.122304 |
| 5 | 10.566880 | 55 | 116.235680 |
| 6 | 12.680256 | 56 | 118.349056 |
| 7 | 14.793632 | 57 | 120.462432 |
| 8 | 16.907008 | 58 | 122.575808 |
| 9 | 19.020384 | 59 | 124.689184 |
| 10 | 21.133760 | 60 | 126.802560 |
| 11 | 23.247136 | 61 | 128.915936 |
| 12 | 25.360512 | 62 | 131.029312 |
| 13 | 27.473888 | 63 | 133.142688 |
| 14 | 29.587264 | 64 | 135.256064 |
| 15 | 31.700640 | 65 | 137.369440 |
| 16 | 33.814016 | 66 | 139.482816 |
| 17 | 35.927392 | 67 | 141.596192 |
| 18 | 38.040768 | 68 | 143.709568 |
| 19 | 40.154144 | 69 | 145.822944 |
| 20 | 42.267520 | 70 | 147.936320 |
| 21 | 44.380896 | 71 | 150.049696 |
| 22 | 46.494272 | 72 | 152.163072 |
| 23 | 48.607648 | 73 | 154.276448 |
| 24 | 50.721024 | 74 | 156.389824 |
| 25 | 52.834400 | 75 | 158.503200 |
| 26 | 54.947776 | 76 | 160.616576 |
| 27 | 57.061152 | 77 | 162.729952 |
| 28 | 59.174528 | 78 | 164.843328 |
| 29 | 61.287904 | 79 | 166.956704 |
| 30 | 63.401280 | 80 | 169.070080 |
| 31 | 65.514656 | 81 | 171.183456 |
| 32 | 67.628032 | 82 | 173.296832 |
| 33 | 69.741408 | 83 | 175.410208 |
| 34 | 71.854784 | 84 | 177.523584 |
| 35 | 73.968160 | 85 | 179.636960 |
| 36 | 76.081536 | 86 | 181.750336 |
| 37 | 78.194912 | 87 | 183.863712 |
| 38 | 80.308288 | 88 | 185.977088 |
| 39 | 82.421664 | 89 | 188.090464 |
| 40 | 84.535040 | 90 | 190.203840 |
| 41 | 86.648416 | 91 | 192.317216 |
| 42 | 88.761792 | 92 | 194.430592 |
| 43 | 90.875168 | 93 | 196.543968 |
| 44 | 92.988544 | 94 | 198.657344 |
| 45 | 95.101920 | 95 | 200.770720 |
| 46 | 97.215296 | 96 | 202.884096 |
| 47 | 99.328672 | 97 | 204.997472 |
| 48 | 101.442048 | 98 | 207.110848 |
| 49 | 103.555424 | 99 | 209.224224 |
| 50 | 105.668800 | 100 | 211.337600 |

## VOLUME
### Quarts (dry) to Cubic Decimeters

| qt | dm³ | qt | dm³ |
|---|---|---|---|
| 1 | 1.101221 | 51 | 56.162271 |
| 2 | 2.202442 | 52 | 57.263492 |
| 3 | 3.303663 | 53 | 58.364713 |
| 4 | 4.404884 | 54 | 59.465934 |
| 5 | 5.506105 | 55 | 60.567155 |
| 6 | 6.607326 | 56 | 61.668376 |
| 7 | 7.708547 | 57 | 62.769597 |
| 8 | 8.809768 | 58 | 63.870818 |
| 9 | 9.910989 | 59 | 64.972039 |
| 10 | 11.012210 | 60 | 66.073260 |
| 11 | 12.113431 | 61 | 67.174481 |
| 12 | 13.214652 | 62 | 68.275702 |
| 13 | 14.315873 | 63 | 69.376923 |
| 14 | 15.417094 | 64 | 70.478144 |
| 15 | 16.518315 | 65 | 71.579365 |
| 16 | 17.619536 | 66 | 72.680586 |
| 17 | 18.720757 | 67 | 73.781807 |
| 18 | 19.821978 | 68 | 74.883028 |
| 19 | 20.923199 | 69 | 75.984249 |
| 20 | 22.024420 | 70 | 77.085470 |
| 21 | 23.125641 | 71 | 78.186691 |
| 22 | 24.226862 | 72 | 79.287912 |
| 23 | 25.328083 | 73 | 80.389133 |
| 24 | 26.429304 | 74 | 81.490354 |
| 25 | 27.530525 | 75 | 82.591575 |
| 26 | 28.631746 | 76 | 83.692796 |
| 27 | 29.732967 | 77 | 84.794017 |
| 28 | 30.834188 | 78 | 85.895238 |
| 29 | 31.935409 | 79 | 86.996459 |
| 30 | 33.036630 | 80 | 88.097680 |
| 31 | 34.137851 | 81 | 89.198901 |
| 32 | 35.239072 | 82 | 90.300122 |
| 33 | 36.340293 | 83 | 91.401343 |
| 34 | 37.441514 | 84 | 92.502564 |
| 35 | 38.542735 | 85 | 93.603785 |
| 36 | 39.643956 | 86 | 94.705006 |
| 37 | 40.745177 | 87 | 95.806227 |
| 38 | 41.846398 | 88 | 96.907448 |
| 39 | 42.947619 | 89 | 98.008669 |
| 40 | 44.048840 | 90 | 99.109890 |
| 41 | 45.150061 | 91 | 100.211111 |
| 42 | 46.251282 | 92 | 101.312332 |
| 43 | 47.352503 | 93 | 102.413553 |
| 44 | 48.453724 | 94 | 103.514774 |
| 45 | 49.554945 | 95 | 104.615995 |
| 46 | 50.656166 | 96 | 105.717216 |
| 47 | 51.757387 | 97 | 106.818437 |
| 48 | 52.858608 | 98 | 107.919658 |
| 49 | 53.959829 | 99 | 109.020879 |
| 50 | 55.061050 | 100 | 110.122100 |

## VOLUME
### Cubic Decimeters to Quarts (dry)

| dm³ | qt | dm³ | qt |
|---|---|---|---|
| 1 | .908082 | 51 | 46.312182 |
| 2 | 1.816164 | 52 | 47.220264 |
| 3 | 2.724246 | 53 | 48.128346 |
| 4 | 3.632328 | 54 | 49.036428 |
| 5 | 4.540410 | 55 | 49.944510 |
| 6 | 5.448492 | 56 | 50.852592 |
| 7 | 6.356574 | 57 | 51.760674 |
| 8 | 7.264656 | 58 | 52.668756 |
| 9 | 8.172738 | 59 | 53.576838 |
| 10 | 9.080820 | 60 | 54.484920 |
| 11 | 9.988902 | 61 | 55.393002 |
| 12 | 10.896984 | 62 | 56.301084 |
| 13 | 11.805066 | 63 | 57.209166 |
| 14 | 12.713148 | 64 | 58.117248 |
| 15 | 13.621230 | 65 | 59.025330 |
| 16 | 14.529312 | 66 | 59.933412 |
| 17 | 15.437394 | 67 | 60.841494 |
| 18 | 16.345476 | 68 | 61.749576 |
| 19 | 17.253558 | 69 | 62.657658 |
| 20 | 18.161640 | 70 | 63.565740 |
| 21 | 19.069722 | 71 | 64.473822 |
| 22 | 19.977804 | 72 | 65.381904 |
| 23 | 20.885886 | 73 | 66.289986 |
| 24 | 21.793968 | 74 | 67.198068 |
| 25 | 22.702050 | 75 | 68.106150 |
| 26 | 23.610132 | 76 | 69.014232 |
| 27 | 24.518214 | 77 | 69.922314 |
| 28 | 25.426296 | 78 | 70.830396 |
| 29 | 26.334378 | 79 | 71.738478 |
| 30 | 27.242460 | 80 | 72.646560 |
| 31 | 28.150542 | 81 | 73.554642 |
| 32 | 29.058624 | 82 | 74.462724 |
| 33 | 29.966706 | 83 | 75.370806 |
| 34 | 30.874788 | 84 | 76.278888 |
| 35 | 31.782870 | 85 | 77.186970 |
| 36 | 32.690952 | 86 | 78.095052 |
| 37 | 33.599034 | 87 | 79.003134 |
| 38 | 34.507116 | 88 | 79.911216 |
| 39 | 35.415198 | 89 | 80.819298 |
| 40 | 36.323280 | 90 | 81.727380 |
| 41 | 37.231362 | 91 | 82.635462 |
| 42 | 38.139444 | 92 | 83.543544 |
| 43 | 39.047526 | 93 | 84.451626 |
| 44 | 39.955608 | 94 | 85.359708 |
| 45 | 40.863690 | 95 | 86.267790 |
| 46 | 41.771772 | 96 | 87.175872 |
| 47 | 42.679854 | 97 | 88.083954 |
| 48 | 43.587936 | 98 | 88.992036 |
| 49 | 44.496018 | 99 | 89.900118 |
| 50 | 45.404100 | 100 | 90.808200 |

## VOLUME
### Quarts (liquid) to Liters

| qt | L | qt | L |
|---|---|---|---|
| 1 | .946353 | 51 | 48.264003 |
| 2 | 1.892706 | 52 | 49.210356 |
| 3 | 2.839059 | 53 | 50.156709 |
| 4 | 3.785412 | 54 | 51.103062 |
| 5 | 4.731765 | 55 | 52.049415 |
| 6 | 5.678118 | 56 | 52.995768 |
| 7 | 6.624471 | 57 | 53.942121 |
| 8 | 7.570824 | 58 | 54.888474 |
| 9 | 8.517177 | 59 | 55.834827 |
| 10 | 9.463530 | 60 | 56.781180 |
| 11 | 10.409883 | 61 | 57.727533 |
| 12 | 11.356236 | 62 | 58.673886 |
| 13 | 12.302589 | 63 | 59.620239 |
| 14 | 13.248942 | 64 | 60.566592 |
| 15 | 14.195295 | 65 | 61.512945 |
| 16 | 15.141648 | 66 | 62.459298 |
| 17 | 16.088001 | 67 | 63.405651 |
| 18 | 17.034354 | 68 | 64.352004 |
| 19 | 17.980707 | 69 | 65.298357 |
| 20 | 18.927060 | 70 | 66.244710 |
| 21 | 19.873413 | 71 | 67.191063 |
| 22 | 20.819766 | 72 | 68.137416 |
| 23 | 21.766119 | 73 | 69.083769 |
| 24 | 22.712472 | 74 | 70.030122 |
| 25 | 23.658825 | 75 | 70.976475 |
| 26 | 24.605178 | 76 | 71.922828 |
| 27 | 25.551531 | 77 | 72.869181 |
| 28 | 26.497884 | 78 | 73.815534 |
| 29 | 27.444237 | 79 | 74.761887 |
| 30 | 28.390590 | 80 | 75.708240 |
| 31 | 29.336943 | 81 | 76.654593 |
| 32 | 30.283296 | 82 | 77.600946 |
| 33 | 31.229649 | 83 | 78.547299 |
| 34 | 32.176002 | 84 | 79.493652 |
| 35 | 33.122355 | 85 | 80.440005 |
| 36 | 34.068708 | 86 | 81.386358 |
| 37 | 35.015061 | 87 | 82.332711 |
| 38 | 35.961414 | 88 | 83.279064 |
| 39 | 36.907767 | 89 | 84.225417 |
| 40 | 37.854120 | 90 | 85.171770 |
| 41 | 38.800473 | 91 | 86.118123 |
| 42 | 39.746826 | 92 | 87.064476 |
| 43 | 40.693179 | 93 | 88.010829 |
| 44 | 41.639532 | 94 | 88.957182 |
| 45 | 42.585885 | 95 | 89.903535 |
| 46 | 43.532238 | 96 | 90.849888 |
| 47 | 44.478591 | 97 | 91.796241 |
| 48 | 45.424944 | 98 | 92.742594 |
| 49 | 46.371297 | 99 | 93.688947 |
| 50 | 47.317650 | 100 | 94.635300 |

## VOLUME
### Liters to Quarts (liquid)

| L | qt | L | qt |
|---|---|---|---|
| 1 | 1.056688 | 51 | 53.891088 |
| 2 | 2.113376 | 52 | 54.947776 |
| 3 | 3.170064 | 53 | 56.004464 |
| 4 | 4.226752 | 54 | 57.061152 |
| 5 | 5.283440 | 55 | 58.117840 |
| 6 | 6.340128 | 56 | 59.174528 |
| 7 | 7.396816 | 57 | 60.231216 |
| 8 | 8.453504 | 58 | 61.287904 |
| 9 | 9.510192 | 59 | 62.344592 |
| 10 | 10.566880 | 60 | 63.401280 |
| 11 | 11.623568 | 61 | 64.457968 |
| 12 | 12.680256 | 62 | 65.514656 |
| 13 | 13.736944 | 63 | 66.571344 |
| 14 | 14.793632 | 64 | 67.628032 |
| 15 | 15.850320 | 65 | 68.684720 |
| 16 | 16.907008 | 66 | 69.741408 |
| 17 | 17.963696 | 67 | 70.798096 |
| 18 | 19.020384 | 68 | 71.854784 |
| 19 | 20.077072 | 69 | 72.911472 |
| 20 | 21.133760 | 70 | 73.968160 |
| 21 | 22.190448 | 71 | 75.024848 |
| 22 | 23.247136 | 72 | 76.081536 |
| 23 | 24.303824 | 73 | 77.138224 |
| 24 | 25.360512 | 74 | 78.194912 |
| 25 | 26.417200 | 75 | 79.251600 |
| 26 | 27.473888 | 76 | 80.308288 |
| 27 | 28.530576 | 77 | 81.364976 |
| 28 | 29.587264 | 78 | 82.421664 |
| 29 | 30.643952 | 79 | 83.478352 |
| 30 | 31.700640 | 80 | 84.535040 |
| 31 | 32.757328 | 81 | 85.591728 |
| 32 | 33.814016 | 82 | 86.648416 |
| 33 | 34.870704 | 83 | 87.705104 |
| 34 | 35.927392 | 84 | 88.761792 |
| 35 | 36.984080 | 85 | 89.818480 |
| 36 | 38.040768 | 86 | 90.875168 |
| 37 | 39.097456 | 87 | 91.931856 |
| 38 | 40.154144 | 88 | 92.988544 |
| 39 | 41.210832 | 89 | 94.045232 |
| 40 | 42.267520 | 90 | 95.101920 |
| 41 | 43.324208 | 91 | 96.158608 |
| 42 | 44.380896 | 92 | 97.215296 |
| 43 | 45.437584 | 93 | 98.271984 |
| 44 | 46.494272 | 94 | 99.328672 |
| 45 | 47.550960 | 95 | 100.385360 |
| 46 | 48.607648 | 96 | 101.442048 |
| 47 | 49.664336 | 97 | 102.498736 |
| 48 | 50.721024 | 98 | 103.555424 |
| 49 | 51.777712 | 99 | 104.612112 |
| 50 | 52.834400 | 100 | 105.668800 |

## VOLUME
### Pecks to Cubic Decimeters

| pk | dm³ | pk | dm³ |
|---|---|---|---|
| 1 | 8.809768 | 51 | 449.298168 |
| 2 | 17.619536 | 52 | 458.107936 |
| 3 | 26.429304 | 53 | 466.917704 |
| 4 | 35.239072 | 54 | 475.727472 |
| 5 | 44.048840 | 55 | 484.537240 |
| 6 | 52.858608 | 56 | 493.347008 |
| 7 | 61.668376 | 57 | 502.156776 |
| 8 | 70.478144 | 58 | 510.966544 |
| 9 | 79.287912 | 59 | 519.776312 |
| 10 | 88.097680 | 60 | 528.586080 |
| 11 | 96.907448 | 61 | 537.395848 |
| 12 | 105.717216 | 62 | 546.205616 |
| 13 | 114.526984 | 63 | 555.015384 |
| 14 | 123.336752 | 64 | 563.825152 |
| 15 | 132.146520 | 65 | 572.634920 |
| 16 | 140.956288 | 66 | 581.444688 |
| 17 | 149.766056 | 67 | 590.254456 |
| 18 | 158.575824 | 68 | 599.064224 |
| 19 | 167.385592 | 69 | 607.873992 |
| 20 | 176.195360 | 70 | 616.683760 |
| 21 | 185.005128 | 71 | 625.493528 |
| 22 | 193.814896 | 72 | 634.303296 |
| 23 | 202.624664 | 73 | 643.113064 |
| 24 | 211.434432 | 74 | 651.922832 |
| 25 | 220.244200 | 75 | 660.732600 |
| 26 | 229.053968 | 76 | 669.542368 |
| 27 | 237.863736 | 77 | 678.352136 |
| 28 | 246.673504 | 78 | 687.161904 |
| 29 | 255.483272 | 79 | 695.971672 |
| 30 | 264.293040 | 80 | 704.781440 |
| 31 | 273.102808 | 81 | 713.591208 |
| 32 | 281.912576 | 82 | 722.400976 |
| 33 | 290.722344 | 83 | 731.210744 |
| 34 | 299.532112 | 84 | 740.020512 |
| 35 | 308.341880 | 85 | 748.830280 |
| 36 | 317.151648 | 86 | 757.640048 |
| 37 | 325.961416 | 87 | 766.449816 |
| 38 | 334.771184 | 88 | 775.259584 |
| 39 | 343.580952 | 89 | 784.069352 |
| 40 | 352.390720 | 90 | 792.879120 |
| 41 | 361.200488 | 91 | 801.688888 |
| 42 | 370.010256 | 92 | 810.498656 |
| 43 | 378.820024 | 93 | 819.308424 |
| 44 | 387.629792 | 94 | 828.118192 |
| 45 | 396.439560 | 95 | 836.927960 |
| 46 | 405.249328 | 96 | 845.737728 |
| 47 | 414.059096 | 97 | 854.547496 |
| 48 | 422.868864 | 98 | 863.357264 |
| 49 | 431.678632 | 99 | 872.167032 |
| 50 | 440.488400 | 100 | 880.976800 |

# VOLUME
## Cubic Decimeters to Pecks

| $dm^3$ | pk | $dm^3$ | pk |
|---|---|---|---|
| 1 | .113510 | 51 | 5.789010 |
| 2 | .227020 | 52 | 5.902520 |
| 3 | .340530 | 53 | 6.016030 |
| 4 | .454040 | 54 | 6.129540 |
| 5 | .567550 | 55 | 6.243050 |
| 6 | .681060 | 56 | 6.356560 |
| 7 | .794570 | 57 | 6.470070 |
| 8 | .908080 | 58 | 6.583580 |
| 9 | 1.021590 | 59 | 6.697090 |
| 10 | 1.135100 | 60 | 6.810600 |
| 11 | 1.248610 | 61 | 6.924110 |
| 12 | 1.362120 | 62 | 7.037620 |
| 13 | 1.475630 | 63 | 7.151130 |
| 14 | 1.589140 | 64 | 7.264640 |
| 15 | 1.702650 | 65 | 7.378150 |
| 16 | 1.816160 | 66 | 7.491660 |
| 17 | 1.929670 | 67 | 7.605170 |
| 18 | 2.043180 | 68 | 7.718680 |
| 19 | 2.156690 | 69 | 7.832190 |
| 20 | 2.270200 | 70 | 7.945700 |
| 21 | 2.383710 | 71 | 8.059210 |
| 22 | 2.497220 | 72 | 8.172720 |
| 23 | 2.610730 | 73 | 8.286230 |
| 24 | 2.724240 | 74 | 8.399740 |
| 25 | 2.837750 | 75 | 8.513250 |
| 26 | 2.951260 | 76 | 8.626760 |
| 27 | 3.064770 | 77 | 8.740270 |
| 28 | 3.178280 | 78 | 8.853780 |
| 29 | 3.291790 | 79 | 8.967290 |
| 30 | 3.405300 | 80 | 9.080800 |
| 31 | 3.518810 | 81 | 9.194310 |
| 32 | 3.632320 | 82 | 9.307820 |
| 33 | 3.745830 | 83 | 9.421330 |
| 34 | 3.859340 | 84 | 9.534840 |
| 35 | 3.972850 | 85 | 9.648350 |
| 36 | 4.086360 | 86 | 9.761860 |
| 37 | 4.199870 | 87 | 9.875370 |
| 38 | 4.313380 | 88 | 9.988880 |
| 39 | 4.426890 | 89 | 10.102390 |
| 40 | 4.540400 | 90 | 10.215900 |
| 41 | 4.653910 | 91 | 10.329410 |
| 42 | 4.767420 | 92 | 10.442920 |
| 43 | 4.880930 | 93 | 10.556430 |
| 44 | 4.994440 | 94 | 10.669940 |
| 45 | 5.107950 | 95 | 10.783450 |
| 46 | 5.221460 | 96 | 10.896960 |
| 47 | 5.334970 | 97 | 11.010470 |
| 48 | 5.448480 | 98 | 11.123980 |
| 49 | 5.561990 | 99 | 11.237490 |
| 50 | 5.675500 | 100 | 11.351000 |

## VOLUME
### Bushels to Cubic Decimeters

| bu | dm³ | bu | dm³ |
|---|---|---|---|
| 1 | 35.238000 | 51 | 1797.138000 |
| 2 | 70.476000 | 52 | 1832.376000 |
| 3 | 105.714000 | 53 | 1867.614000 |
| 4 | 140.952000 | 54 | 1902.852000 |
| 5 | 176.190000 | 55 | 1938.090000 |
| 6 | 211.428000 | 56 | 1973.328000 |
| 7 | 246.666000 | 57 | 2008.566000 |
| 8 | 281.904000 | 58 | 2043.804000 |
| 9 | 317.142000 | 59 | 2079.042000 |
| 10 | 352.380000 | 60 | 2114.280000 |
| 11 | 387.618000 | 61 | 2149.518000 |
| 12 | 422.856000 | 62 | 2184.756000 |
| 13 | 458.094000 | 63 | 2219.994000 |
| 14 | 493.332000 | 64 | 2255.232000 |
| 15 | 528.570000 | 65 | 2290.470000 |
| 16 | 563.808000 | 66 | 2325.708000 |
| 17 | 599.046000 | 67 | 2360.946000 |
| 18 | 634.284000 | 68 | 2396.184000 |
| 19 | 669.522000 | 69 | 2431.422000 |
| 20 | 704.760000 | 70 | 2466.660000 |
| 21 | 739.998000 | 71 | 2501.898000 |
| 22 | 775.236000 | 72 | 2537.136000 |
| 23 | 810.474000 | 73 | 2572.374000 |
| 24 | 845.712000 | 74 | 2607.612000 |
| 25 | 880.950000 | 75 | 2642.850000 |
| 26 | 916.188000 | 76 | 2678.088000 |
| 27 | 951.426000 | 77 | 2713.326000 |
| 28 | 986.664000 | 78 | 2748.564000 |
| 29 | 1021.902000 | 79 | 2783.802000 |
| 30 | 1057.140000 | 80 | 2819.040000 |
| 31 | 1092.378000 | 81 | 2854.278000 |
| 32 | 1127.616000 | 82 | 2889.516000 |
| 33 | 1162.854000 | 83 | 2924.754000 |
| 34 | 1198.092000 | 84 | 2959.992000 |
| 35 | 1233.330000 | 85 | 2995.230000 |
| 36 | 1268.568000 | 86 | 3030.468000 |
| 37 | 1303.806000 | 87 | 3065.706000 |
| 38 | 1339.044000 | 88 | 3100.944000 |
| 39 | 1374.282000 | 89 | 3136.182000 |
| 40 | 1409.520000 | 90 | 3171.420000 |
| 41 | 1444.758000 | 91 | 3206.658000 |
| 42 | 1479.996000 | 92 | 3241.896000 |
| 43 | 1515.234000 | 93 | 3277.134000 |
| 44 | 1550.472000 | 94 | 3312.372000 |
| 45 | 1585.710000 | 95 | 3347.610000 |
| 46 | 1620.948000 | 96 | 3382.848000 |
| 47 | 1656.186000 | 97 | 3418.086000 |
| 48 | 1691.424000 | 98 | 3453.324000 |
| 49 | 1726.662000 | 99 | 3488.562000 |
| 50 | 1761.900000 | 100 | 3523.800000 |

## VOLUME
### Cubic Decimeters to Bushels

| dm³ | bu | dm³ | bu |
|---|---|---|---|
| 1 | .028378 | 51 | 1.447278 |
| 2 | .056756 | 52 | 1.475656 |
| 3 | .085134 | 53 | 1.504034 |
| 4 | .113512 | 54 | 1.532412 |
| 5 | .141890 | 55 | 1.560790 |
| 6 | .170268 | 56 | 1.589168 |
| 7 | .198646 | 57 | 1.617546 |
| 8 | .227024 | 58 | 1.645924 |
| 9 | .255402 | 59 | 1.674302 |
| 10 | .283780 | 60 | 1.702680 |
| 11 | .312158 | 61 | 1.731058 |
| 12 | .340536 | 62 | 1.759436 |
| 13 | .368914 | 63 | 1.787814 |
| 14 | .397292 | 64 | 1.816192 |
| 15 | .425670 | 65 | 1.844570 |
| 16 | .454048 | 66 | 1.872948 |
| 17 | .482426 | 67 | 1.901326 |
| 18 | .510804 | 68 | 1.929704 |
| 19 | .539182 | 69 | 1.958082 |
| 20 | .567560 | 70 | 1.986460 |
| 21 | .595938 | 71 | 2.014838 |
| 22 | .624316 | 72 | 2.043216 |
| 23 | .652694 | 73 | 2.071594 |
| 24 | .681072 | 74 | 2.099972 |
| 25 | .709450 | 75 | 2.128350 |
| 26 | .737828 | 76 | 2.156728 |
| 27 | .766206 | 77 | 2.185106 |
| 28 | .794584 | 78 | 2.213484 |
| 29 | .822962 | 79 | 2.241862 |
| 30 | .851340 | 80 | 2.270240 |
| 31 | .879718 | 81 | 2.298618 |
| 32 | .908096 | 82 | 2.326996 |
| 33 | .936474 | 83 | 2.355374 |
| 34 | .964852 | 84 | 2.383752 |
| 35 | .993230 | 85 | 2.412130 |
| 36 | 1.021608 | 86 | 2.440508 |
| 37 | 1.049986 | 87 | 2.468886 |
| 38 | 1.078364 | 88 | 2.497264 |
| 39 | 1.106742 | 89 | 2.525642 |
| 40 | 1.135120 | 90 | 2.554020 |
| 41 | 1.163498 | 91 | 2.582398 |
| 42 | 1.191876 | 92 | 2.610776 |
| 43 | 1.220254 | 93 | 2.639154 |
| 44 | 1.248632 | 94 | 2.667532 |
| 45 | 1.277010 | 95 | 2.695910 |
| 46 | 1.305388 | 96 | 2.724288 |
| 47 | 1.333766 | 97 | 2.752666 |
| 48 | 1.362144 | 98 | 2.781044 |
| 49 | 1.390522 | 99 | 2.809422 |
| 50 | 1.418900 | 100 | 2.837800 |

## VOLUME
### Gallons to Liters

| gal | L | gal | L |
|---|---|---|---|
| 1 | 3.785412 | 51 | 193.056012 |
| 2 | 7.570824 | 52 | 196.841424 |
| 3 | 11.356236 | 53 | 200.626836 |
| 4 | 15.141648 | 54 | 204.412248 |
| 5 | 18.927060 | 55 | 208.197660 |
| 6 | 22.712472 | 56 | 211.983072 |
| 7 | 26.497884 | 57 | 215.768484 |
| 8 | 30.283296 | 58 | 219.553896 |
| 9 | 34.068708 | 59 | 223.339308 |
| 10 | 37.854120 | 60 | 227.124720 |
| 11 | 41.639532 | 61 | 230.910132 |
| 12 | 45.424944 | 62 | 234.695544 |
| 13 | 49.210356 | 63 | 238.480956 |
| 14 | 52.995768 | 64 | 242.266368 |
| 15 | 56.781180 | 65 | 246.051780 |
| 16 | 60.566592 | 66 | 249.837192 |
| 17 | 64.352004 | 67 | 253.622604 |
| 18 | 68.137416 | 68 | 257.408016 |
| 19 | 71.922828 | 69 | 261.193428 |
| 20 | 75.708240 | 70 | 264.978840 |
| 21 | 79.493652 | 71 | 268.764252 |
| 22 | 83.279064 | 72 | 272.549664 |
| 23 | 87.064476 | 73 | 276.335076 |
| 24 | 90.849888 | 74 | 280.120488 |
| 25 | 94.635300 | 75 | 283.905900 |
| 26 | 98.420712 | 76 | 287.691312 |
| 27 | 102.206124 | 77 | 291.476724 |
| 28 | 105.991536 | 78 | 295.262136 |
| 29 | 109.776948 | 79 | 299.047548 |
| 30 | 113.562360 | 80 | 302.832960 |
| 31 | 117.347772 | 81 | 306.618372 |
| 32 | 121.133184 | 82 | 310.403784 |
| 33 | 124.918596 | 83 | 314.189196 |
| 34 | 128.704008 | 84 | 317.974608 |
| 35 | 132.489420 | 85 | 321.760020 |
| 36 | 136.274832 | 86 | 325.545432 |
| 37 | 140.060244 | 87 | 329.330844 |
| 38 | 143.845656 | 88 | 333.116256 |
| 39 | 147.631068 | 89 | 336.901668 |
| 40 | 151.416480 | 90 | 340.687080 |
| 41 | 155.201892 | 91 | 344.472492 |
| 42 | 158.987304 | 92 | 348.257904 |
| 43 | 162.772716 | 93 | 352.043316 |
| 44 | 166.558128 | 94 | 355.828728 |
| 45 | 170.343540 | 95 | 359.614140 |
| 46 | 174.128952 | 96 | 363.399552 |
| 47 | 177.914364 | 97 | 367.184964 |
| 48 | 181.699776 | 98 | 370.970376 |
| 49 | 185.485188 | 99 | 374.755788 |
| 50 | 189.270600 | 100 | 378.541200 |

## VOLUME
### Liters to Gallons

| L | gal | L | gal |
|---|---|---|---|
| 1 | .264172 | 51 | 13.472772 |
| 2 | .528344 | 52 | 13.736944 |
| 3 | .792516 | 53 | 14.001116 |
| 4 | 1.056688 | 54 | 14.265288 |
| 5 | 1.320860 | 55 | 14.529460 |
| 6 | 1.585032 | 56 | 14.793632 |
| 7 | 1.849204 | 57 | 15.057804 |
| 8 | 2.113376 | 58 | 15.321976 |
| 9 | 2.377548 | 59 | 15.586148 |
| 10 | 2.641720 | 60 | 15.850320 |
| 11 | 2.905892 | 61 | 16.114492 |
| 12 | 3.170064 | 62 | 16.378664 |
| 13 | 3.434236 | 63 | 16.642836 |
| 14 | 3.698408 | 64 | 16.907008 |
| 15 | 3.962580 | 65 | 17.171180 |
| 16 | 4.226752 | 66 | 17.435352 |
| 17 | 4.490924 | 67 | 17.699524 |
| 18 | 4.755096 | 68 | 17.963696 |
| 19 | 5.019268 | 69 | 18.227868 |
| 20 | 5.283440 | 70 | 18.492040 |
| 21 | 5.547612 | 71 | 18.756212 |
| 22 | 5.811784 | 72 | 19.020384 |
| 23 | 6.075956 | 73 | 19.284556 |
| 24 | 6.340128 | 74 | 19.548728 |
| 25 | 6.604300 | 75 | 19.812900 |
| 26 | 6.868472 | 76 | 20.077072 |
| 27 | 7.132644 | 77 | 20.341244 |
| 28 | 7.396816 | 78 | 20.605416 |
| 29 | 7.660988 | 79 | 20.869588 |
| 30 | 7.925160 | 80 | 21.133760 |
| 31 | 8.189332 | 81 | 21.397932 |
| 32 | 8.453504 | 82 | 21.662104 |
| 33 | 8.717676 | 83 | 21.926276 |
| 34 | 8.981848 | 84 | 22.190448 |
| 35 | 9.246020 | 85 | 22.454620 |
| 36 | 9.510192 | 86 | 22.718792 |
| 37 | 9.774364 | 87 | 22.982964 |
| 38 | 10.038536 | 88 | 23.247136 |
| 39 | 10.302708 | 89 | 23.511308 |
| 40 | 10.566880 | 90 | 23.775480 |
| 41 | 10.831052 | 91 | 24.039652 |
| 42 | 11.095224 | 92 | 24.303824 |
| 43 | 11.359396 | 93 | 24.567996 |
| 44 | 11.623568 | 94 | 24.832168 |
| 45 | 11.887740 | 95 | 25.096340 |
| 46 | 12.151912 | 96 | 25.360512 |
| 47 | 12.416084 | 97 | 25.624684 |
| 48 | 12.680256 | 98 | 25.888856 |
| 49 | 12.944428 | 99 | 26.153028 |
| 50 | 13.208600 | 100 | 26.417200 |

## VOLUME
### Acre Feet to Cubic Meters

| ac ft | m³ | ac ft | m³ |
|---|---|---|---|
| 1 | 1233.482000 | 51 | 62907.582000 |
| 2 | 2466.964000 | 52 | 64141.064000 |
| 3 | 3700.446000 | 53 | 65374.546000 |
| 4 | 4933.928000 | 54 | 66608.028000 |
| 5 | 6167.410000 | 55 | 67841.510000 |
| 6 | 7400.892000 | 56 | 69074.992000 |
| 7 | 8634.374000 | 57 | 70308.474000 |
| 8 | 9867.856000 | 58 | 71541.956000 |
| 9 | 11101.338000 | 59 | 72775.438000 |
| 10 | 12334.820000 | 60 | 74008.920000 |
| 11 | 13568.302000 | 61 | 75242.402000 |
| 12 | 14801.784000 | 62 | 76475.884000 |
| 13 | 16035.266000 | 63 | 77709.366000 |
| 14 | 17268.748000 | 64 | 78942.848000 |
| 15 | 18502.230000 | 65 | 80176.330000 |
| 16 | 19735.712000 | 66 | 81409.812000 |
| 17 | 20969.194000 | 67 | 82643.294000 |
| 18 | 22202.676000 | 68 | 83876.776000 |
| 19 | 23436.158000 | 69 | 85110.258000 |
| 20 | 24669.640000 | 70 | 86343.740000 |
| 21 | 25903.122000 | 71 | 87577.222000 |
| 22 | 27136.604000 | 72 | 88810.704000 |
| 23 | 28370.086000 | 73 | 90044.186000 |
| 24 | 29603.568000 | 74 | 91277.668000 |
| 25 | 30837.050000 | 75 | 92511.150000 |
| 26 | 32070.532000 | 76 | 93744.632000 |
| 27 | 33304.014000 | 77 | 94978.114000 |
| 28 | 34537.496000 | 78 | 96211.596000 |
| 29 | 35770.978000 | 79 | 97445.078000 |
| 30 | 37004.460000 | 80 | 98678.560000 |
| 31 | 38237.942000 | 81 | 99912.042000 |
| 32 | 39471.424000 | 82 | 101145.524000 |
| 33 | 40704.906000 | 83 | 102379.006000 |
| 34 | 41938.388000 | 84 | 103612.488000 |
| 35 | 43171.870000 | 85 | 104845.970000 |
| 36 | 44405.352000 | 86 | 106079.452000 |
| 37 | 45638.834000 | 87 | 107312.934000 |
| 38 | 46872.316000 | 88 | 108546.416000 |
| 39 | 48105.798000 | 89 | 109779.898000 |
| 40 | 49339.280000 | 90 | 111013.380000 |
| 41 | 50572.762000 | 91 | 112246.862000 |
| 42 | 51806.244000 | 92 | 113480.344000 |
| 43 | 53039.726000 | 93 | 114713.826000 |
| 44 | 54273.208000 | 94 | 115947.308000 |
| 45 | 55506.690000 | 95 | 117180.790000 |
| 46 | 56740.172000 | 96 | 118414.272000 |
| 47 | 57973.654000 | 97 | 119647.754000 |
| 48 | 59207.136000 | 98 | 120881.236000 |
| 49 | 60440.618000 | 99 | 122114.718000 |
| 50 | 61674.100000 | 100 | 123348.200000 |

## VOLUME
### Cubic Meters to Acre Feet

| $m^3$ | ac ft | $m^3$ | ac ft |
|---|---|---|---|
| 1 | .000810 | 51 | .041310 |
| 2 | .001620 | 52 | .042120 |
| 3 | .002430 | 53 | .042930 |
| 4 | ,003240 | 54 | .043740 |
| 5 | .004050 | 55 | .044550 |
| 6 | .004860 | 56 | .045360 |
| 7 | .005670 | 57 | .046170 |
| 8 | .006480 | 58 | .046980 |
| 9 | .007290 | 59 | .047790 |
| 10 | .008100 | 60 | .048600 |
| 11 | .008910 | 61 | .049410 |
| 12 | .009720 | 62 | .050220 |
| 13 | .010530 | 63 | .051030 |
| 14 | .011340 | 64 | .051840 |
| 15 | .012150 | 65 | .052650 |
| 16 | .012960 | 66 | .053460 |
| 17 | .013770 | 67 | .054270 |
| 18 | .014580 | 68 | .055080 |
| 19 | .015390 | 69 | .055890 |
| 20 | .016200 | 70 | .056700 |
| 21 | .017010 | 71 | .057510 |
| 22 | .017820 | 72 | .058320 |
| 23 | .018630 | 73 | .059130 |
| 24 | .019440 | 74 | .059940 |
| 25 | .020250 | 75 | .060750 |
| 26 | .021060 | 76 | .061560 |
| 27 | .021870 | 77 | .062370 |
| 28 | .022680 | 78 | .063180 |
| 29 | .023490 | 79 | .063990 |
| 30 | .024300 | 80 | .064800 |
| 31 | .025110 | 81 | .065610 |
| 32 | .025920 | 82 | .066420 |
| 33 | .026730 | 83 | .067230 |
| 34 | .027540 | 84 | .068040 |
| 35 | .028350 | 85 | .068850 |
| 36 | .029160 | 86 | .069660 |
| 37 | .029970 | 87 | .070470 |
| 38 | .030780 | 88 | .071280 |
| 39 | .031590 | 89 | .072090 |
| 40 | .032400 | 90 | .072900 |
| 41 | .033210 | 91 | .073710 |
| 42 | .034020 | 92 | .074520 |
| 43 | .034830 | 93 | .075330 |
| 44 | .035640 | 94 | .076140 |
| 45 | .036450 | 95 | .076950 |
| 46 | .037260 | 96 | .077760 |
| 47 | .038070 | 97 | .078570 |
| 48 | .038880 | 98 | .079380 |
| 49 | .039690 | 99 | .080190 |
| 50 | .040500 | 100 | .081000 |

## VOLUME
### Board Feet to Cubic Meters

| bd ft | m³ | bd ft | m³ |
|---|---|---|---|
| 1 | .002359 | 51 | .120309 |
| 2 | .004718 | 52 | .122668 |
| 3 | .007077 | 53 | .125027 |
| 4 | .009436 | 54 | .127386 |
| 5 | .011795 | 55 | .129745 |
| 6 | .014154 | 56 | .132104 |
| 7 | .016513 | 57 | .134463 |
| 8 | .018872 | 58 | .136822 |
| 9 | .021231 | 59 | .139181 |
| 10 | .023590 | 60 | .141540 |
| 11 | .025949 | 61 | .143899 |
| 12 | .028308 | 62 | .146258 |
| 13 | .030667 | 63 | .148617 |
| 14 | .033026 | 64 | .150976 |
| 15 | .035385 | 65 | .153335 |
| 16 | .037744 | 66 | .155694 |
| 17 | .040103 | 67 | .158053 |
| 18 | .042462 | 68 | .160412 |
| 19 | .044821 | 69 | .162771 |
| 20 | .047180 | 70 | .165130 |
| 21 | .049539 | 71 | .167489 |
| 22 | .051898 | 72 | .169848 |
| 23 | .054257 | 73 | .172207 |
| 24 | .056616 | 74 | .174566 |
| 25 | .058975 | 75 | .176925 |
| 26 | .061334 | 76 | .179284 |
| 27 | .063693 | 77 | .181643 |
| 28 | .066052 | 78 | .184002 |
| 29 | .068411 | 79 | .186361 |
| 30 | .070770 | 80 | .188720 |
| 31 | .073129 | 81 | .191079 |
| 32 | .075488 | 82 | .193438 |
| 33 | .077847 | 83 | .195797 |
| 34 | .080206 | 84 | .198156 |
| 35 | .082565 | 85 | .200515 |
| 36 | .084924 | 86 | .202874 |
| 37 | .087283 | 87 | .205233 |
| 38 | .089642 | 88 | .207592 |
| 39 | .092001 | 89 | .209951 |
| 40 | .094360 | 90 | .212310 |
| 41 | .096719 | 91 | .214669 |
| 42 | .099078 | 92 | .217028 |
| 43 | .101437 | 93 | .219387 |
| 44 | .103796 | 94 | .221746 |
| 45 | .106155 | 95 | .224105 |
| 46 | .108514 | 96 | .226464 |
| 47 | .110873 | 97 | .228823 |
| 48 | .113232 | 98 | .231182 |
| 49 | .115591 | 99 | .233541 |
| 50 | .117950 | 100 | .235900 |

# VOLUME
## Cubic Meters to Board Feet

| m³ | bd ft | m³ | bd ft |
|---|---|---|---|
| 1 | 423.776039 | 51 | 21612.577989 |
| 2 | 847.552078 | 52 | 22036.354028 |
| 3 | 1271.328117 | 53 | 22460.130067 |
| 4 | 1695.104156 | 54 | 22883.906106 |
| 5 | 2118.880195 | 55 | 23307.682145 |
| 6 | 2542.656234 | 56 | 23731.458184 |
| 7 | 2966.432273 | 57 | 24155.234223 |
| 8 | 3390.208312 | 58 | 24579.010262 |
| 9 | 3813.984351 | 59 | 25002.786301 |
| 10 | 4237.760390 | 60 | 25426.562340 |
| 11 | 4661.536429 | 61 | 25850.338379 |
| 12 | 5085.312468 | 62 | 26274.114418 |
| 13 | 5509.088507 | 63 | 26697.890457 |
| 14 | 5932.864546 | 64 | 27121.666496 |
| 15 | 6356.640585 | 65 | 27545.442535 |
| 16 | 6780.416624 | 66 | 27969.218574 |
| 17 | 7204.192663 | 67 | 28392.994613 |
| 18 | 7627.968702 | 68 | 28816.770652 |
| 19 | 8051.744741 | 69 | 29240.546691 |
| 20 | 8475.520780 | 70 | 29664.322730 |
| 21 | 8899.296819 | 71 | 30088.098769 |
| 22 | 9323.072858 | 72 | 30511.874808 |
| 23 | 9746.848897 | 73 | 30935.650847 |
| 24 | 10170.624936 | 74 | 31359.426886 |
| 25 | 10594.400975 | 75 | 31783.202925 |
| 26 | 11018.177014 | 76 | 32206.978964 |
| 27 | 11441.953053 | 77 | 32630.755003 |
| 28 | 11865.729092 | 78 | 33054.531042 |
| 29 | 12289.505131 | 79 | 33478.307081 |
| 30 | 12713.281170 | 80 | 33902.083120 |
| 31 | 13137.057209 | 81 | 34325.859159 |
| 32 | 13560.833248 | 82 | 34749.635198 |
| 33 | 13984.609287 | 83 | 35173.411237 |
| 34 | 14408.385326 | 84 | 35597.187276 |
| 35 | 14832.161365 | 85 | 36020.963315 |
| 36 | 15255.937404 | 86 | 36444.739354 |
| 37 | 15679.713443 | 87 | 36868.515393 |
| 38 | 16103.489482 | 88 | 37292.291432 |
| 39 | 16527.265521 | 89 | 37716.067471 |
| 40 | 16951.041560 | 90 | 38139.843510 |
| 41 | 17374.817599 | 91 | 38563.619549 |
| 42 | 17798.593638 | 92 | 38987.395588 |
| 43 | 18222.369677 | 93 | 39411.171627 |
| 44 | 18646.145716 | 94 | 39834.947666 |
| 45 | 19069.921755 | 95 | 40258.723705 |
| 46 | 19493.697794 | 96 | 40682.499744 |
| 47 | 19917.473833 | 97 | 41106.275783 |
| 48 | 20341.249872 | 98 | 41530.051822 |
| 49 | 20765.025911 | 99 | 41953.827861 |
| 50 | 21188.801950 | 100 | 42377.603900 |

## VOLUME
### Cords to Cubic Meters

| cord | m³ | cord | m³ |
|---|---|---|---|
| 1 | 3.624556 | 51 | 184.852356 |
| 2 | 7.249112 | 52 | 188.476912 |
| 3 | 10.873668 | 53 | 192.101468 |
| 4 | 14.498224 | 54 | 195.726024 |
| 5 | 18.122780 | 55 | 199.350580 |
| 6 | 21.747336 | 56 | 202.975136 |
| 7 | 25.371892 | 57 | 206.599692 |
| 8 | 28.996448 | 58 | 210.224248 |
| 9 | 32.621004 | 59 | 213.848804 |
| 10 | 36.245560 | 60 | 217.473360 |
| 11 | 39.870116 | 61 | 221.097916 |
| 12 | 43.494672 | 62 | 224.722472 |
| 13 | 47.119228 | 63 | 228.347028 |
| 14 | 50.743784 | 64 | 231.971584 |
| 15 | 54.368340 | 65 | 235.596140 |
| 16 | 57.992896 | 66 | 239.220696 |
| 17 | 61.617452 | 67 | 242.845252 |
| 18 | 65.242008 | 68 | 246.469808 |
| 19 | 68.866564 | 69 | 250.094364 |
| 20 | 72.491120 | 70 | 253.718920 |
| 21 | 76.115676 | 71 | 257.343476 |
| 22 | 79.740232 | 72 | 260.968032 |
| 23 | 83.364788 | 73 | 264.592588 |
| 24 | 86.989344 | 74 | 268.217144 |
| 25 | 90.613900 | 75 | 271.841700 |
| 26 | 94.238456 | 76 | 275.466256 |
| 27 | 97.863012 | 77 | 279.090812 |
| 28 | 101.487568 | 78 | 282.715368 |
| 29 | 105.112124 | 79 | 286.339924 |
| 30 | 108.736680 | 80 | 289.964480 |
| 31 | 112.361236 | 81 | 293.589036 |
| 32 | 115.985792 | 82 | 297.213592 |
| 33 | 119.610348 | 83 | 300.838148 |
| 34 | 123.234904 | 84 | 304.462704 |
| 35 | 126.859460 | 85 | 308.087260 |
| 36 | 130.484016 | 86 | 311.711816 |
| 37 | 134.108572 | 87 | 315.336372 |
| 38 | 137.733128 | 88 | 318.960928 |
| 39 | 141.357684 | 89 | 322.585484 |
| 40 | 144.982240 | 90 | 326.210040 |
| 41 | 148.606796 | 91 | 329.834596 |
| 42 | 152.231352 | 92 | 333.459152 |
| 43 | 155.855908 | 93 | 337.083708 |
| 44 | 159.480464 | 94 | 340.708264 |
| 45 | 163.105020 | 95 | 344.332820 |
| 46 | 166.729576 | 96 | 347.957376 |
| 47 | 170.354132 | 97 | 351.581932 |
| 48 | 173.978688 | 98 | 355.206488 |
| 49 | 177.603244 | 99 | 358.831044 |
| 50 | 181.227800 | 100 | 362.455600 |

# VOLUME
## Cubic Meters to Cords

| m³ | cord | m³ | cord |
|---|---|---|---|
| 1 | .275896 | 51 | 14.070696 |
| 2 | .551792 | 52 | 14.346592 |
| 3 | .827688 | 53 | 14.622488 |
| 4 | 1.103584 | 54 | 14.898384 |
| 5 | 1.379480 | 55 | 15.174280 |
| 6 | 1.655376 | 56 | 15.450176 |
| 7 | 1.931272 | 57 | 15.726072 |
| 8 | 2.207168 | 58 | 16.001968 |
| 9 | 2.483064 | 59 | 16.277864 |
| 10 | 2.758960 | 60 | 16.553760 |
| 11 | 3.034856 | 61 | 16.829656 |
| 12 | 3.310752 | 62 | 17.105552 |
| 13 | 3.586648 | 63 | 17.381448 |
| 14 | 3.862544 | 64 | 17.657344 |
| 15 | 4.138440 | 65 | 17.933240 |
| 16 | 4.414336 | 66 | 18.209136 |
| 17 | 4.690232 | 67 | 18.485032 |
| 18 | 4.966128 | 68 | 18.760928 |
| 19 | 5.242024 | 69 | 19.036824 |
| 20 | 5.517920 | 70 | 19.312720 |
| 21 | 5.793816 | 71 | 19.588616 |
| 22 | 6.069712 | 72 | 19.864512 |
| 23 | 6.345608 | 73 | 20.140408 |
| 24 | 6.621504 | 74 | 20.416304 |
| 25 | 6.897400 | 75 | 20.692200 |
| 26 | 7.173296 | 76 | 20.968096 |
| 27 | 7.449192 | 77 | 21.243992 |
| 28 | 7.725088 | 78 | 21.519888 |
| 29 | 8.000984 | 79 | 21.795784 |
| 30 | 8.276880 | 80 | 22.071680 |
| 31 | 8.552776 | 81 | 22.347576 |
| 32 | 8.828672 | 82 | 22.623472 |
| 33 | 9.104568 | 83 | 22.899368 |
| 34 | 9.380464 | 84 | 23.175264 |
| 35 | 9.656360 | 85 | 23.451160 |
| 36 | 9.932256 | 86 | 23.727056 |
| 37 | 10.208152 | 87 | 24.002952 |
| 38 | 10.484048 | 88 | 24.278848 |
| 39 | 10.759944 | 89 | 24.554744 |
| 40 | 11.035840 | 90 | 24.830640 |
| 41 | 11.311736 | 91 | 25.106536 |
| 42 | 11.587632 | 92 | 25.382432 |
| 43 | 11.863528 | 93 | 25.658328 |
| 44 | 12.139424 | 94 | 25.934224 |
| 45 | 12.415320 | 95 | 26.210120 |
| 46 | 12.691216 | 96 | 26.486016 |
| 47 | 12.967112 | 97 | 26.761912 |
| 48 | 13.243008 | 98 | 27.037808 |
| 49 | 13.518904 | 99 | 27.313704 |
| 50 | 13.794800 | 100 | 27.589600 |

## WEIGHT
### Grains to Milligrams

| gr | mg | gr | mg |
|---|---|---|---|
| 1 | 64.798910 | 51 | 3304.744410 |
| 2 | 129.597820 | 52 | 3369.543320 |
| 3 | 194.396730 | 53 | 3434.342230 |
| 4 | 259.195640 | 54 | 3499.141140 |
| 5 | 323.994550 | 55 | 3563.940050 |
| 6 | 388.793460 | 56 | 3628.738960 |
| 7 | 453.592370 | 57 | 3693.537870 |
| 8 | 518.391280 | 58 | 3758.336780 |
| 9 | 583.190190 | 59 | 3823.135690 |
| 10 | 647.989100 | 60 | 3887.934600 |
| 11 | 712.788010 | 61 | 3952.733510 |
| 12 | 777.586920 | 62 | 4017.532420 |
| 13 | 842.385830 | 63 | 4082.331330 |
| 14 | 907.184740 | 64 | 4147.130240 |
| 15 | 971.983650 | 65 | 4211.929150 |
| 16 | 1036.782560 | 66 | 4276.728060 |
| 17 | 1101.581470 | 67 | 4341.526970 |
| 18 | 1166.380380 | 68 | 4406.325880 |
| 19 | 1231.179290 | 69 | 4471.124790 |
| 20 | 1295.978200 | 70 | 4535.923700 |
| 21 | 1360.777110 | 71 | 4600.722610 |
| 22 | 1425.576020 | 72 | 4665.521520 |
| 23 | 1490.374930 | 73 | 4730.320430 |
| 24 | 1555.173840 | 74 | 4795.119340 |
| 25 | 1619.972750 | 75 | 4859.918250 |
| 26 | 1684.771660 | 76 | 4924.717160 |
| 27 | 1749.570570 | 77 | 4989.516070 |
| 28 | 1814.369480 | 78 | 5054.314980 |
| 29 | 1879.168390 | 79 | 5119.113890 |
| 30 | 1943.967300 | 80 | 5183.912800 |
| 31 | 2008.766210 | 81 | 5248.711710 |
| 32 | 2073.565120 | 82 | 5313.510620 |
| 33 | 2138.364030 | 83 | 5378.309530 |
| 34 | 2203.162940 | 84 | 5443.108440 |
| 35 | 2267.961850 | 85 | 5507.907350 |
| 36 | 2332.760760 | 86 | 5572.706260 |
| 37 | 2397.559670 | 87 | 5637.505170 |
| 38 | 2462.358580 | 88 | 5702.304080 |
| 39 | 2527.157490 | 89 | 5767.102990 |
| 40 | 2591.956400 | 90 | 5831.901900 |
| 41 | 2656.755310 | 91 | 5896.700810 |
| 42 | 2721.554220 | 92 | 5961.499720 |
| 43 | 2786.353130 | 93 | 6026.298630 |
| 44 | 2851.152040 | 94 | 6091.097540 |
| 45 | 2915.950950 | 95 | 6155.896450 |
| 46 | 2980.749860 | 96 | 6220.695360 |
| 47 | 3045.548770 | 97 | 6285.494270 |
| 48 | 3110.347680 | 98 | 6350.293180 |
| 49 | 3175.146590 | 99 | 6415.092090 |
| 50 | 3239.945500 | 100 | 6479.891000 |

## WEIGHT
### Milligrams to Grains

| mg | gr | mg | gr |
|---|---|---|---|
| 1 | .015432 | 51 | .787032 |
| 2 | .030864 | 52 | .802464 |
| 3 | .046296 | 53 | .817896 |
| 4 | .061728 | 54 | .833328 |
| 5 | .077160 | 55 | .848760 |
| 6 | .092592 | 56 | .864192 |
| 7 | .108024 | 57 | .879624 |
| 8 | .123456 | 58 | .895056 |
| 9 | .138888 | 59 | .910488 |
| 10 | .154320 | 60 | .925920 |
| 11 | .169752 | 61 | .941352 |
| 12 | .185184 | 62 | .956784 |
| 13 | .200616 | 63 | .972216 |
| 14 | .216048 | 64 | .987648 |
| 15 | .231480 | 65 | 1.003080 |
| 16 | .246912 | 66 | 1.018512 |
| 17 | .262344 | 67 | 1.033944 |
| 18 | .277776 | 68 | 1.049376 |
| 19 | .293208 | 69 | 1.064808 |
| 20 | .308640 | 70 | 1.080240 |
| 21 | .324072 | 71 | 1.095672 |
| 22 | .339504 | 72 | 1.111104 |
| 23 | .354936 | 73 | 1.126536 |
| 24 | .370368 | 74 | 1.141968 |
| 25 | .385800 | 75 | 1.157400 |
| 26 | .401232 | 76 | 1.172832 |
| 27 | .416664 | 77 | 1.188264 |
| 28 | .432096 | 78 | 1.203696 |
| 29 | .447528 | 79 | 1.219128 |
| 30 | .462960 | 80 | 1.234560 |
| 31 | .478392 | 81 | 1.249992 |
| 32 | .493824 | 82 | 1.265424 |
| 33 | .509256 | 83 | 1.280856 |
| 34 | .524688 | 84 | 1.296288 |
| 35 | .540120 | 85 | 1.311720 |
| 36 | .555552 | 86 | 1.327152 |
| 37 | .570984 | 87 | 1.342584 |
| 38 | .586416 | 88 | 1.358016 |
| 39 | .601848 | 89 | 1.373448 |
| 40 | .617280 | 90 | 1.388880 |
| 41 | .632712 | 91 | 1.404312 |
| 42 | .648144 | 92 | 1.419744 |
| 43 | .663576 | 93 | 1.435176 |
| 44 | .679008 | 94 | 1.450608 |
| 45 | .694440 | 95 | 1.466040 |
| 46 | .709872 | 96 | 1.481472 |
| 47 | .725304 | 97 | 1.496904 |
| 48 | .740736 | 98 | 1.512336 |
| 49 | .756168 | 99 | 1.527768 |
| 50 | .771600 | 100 | 1.543200 |

## WEIGHT

### Drams (apothecaries') to Grams

| dr ap | g | dr ap | g |
|---|---|---|---|
| 1 | 3.887935 | 51 | 198.284685 |
| 2 | 7.775870 | 52 | 202.172620 |
| 3 | 11.663805 | 53 | 206.060555 |
| 4 | 15.551740 | 54 | 209.948490 |
| 5 | 19.439675 | 55 | 213.836425 |
| 6 | 23.327610 | 56 | 217.724360 |
| 7 | 27.215545 | 57 | 221.612295 |
| 8 | 31.103480 | 58 | 225.500230 |
| 9 | 34.991415 | 59 | 229.388165 |
| 10 | 38.879350 | 60 | 233.276100 |
| 11 | 42.767285 | 61 | 237.164035 |
| 12 | 46.655220 | 62 | 241.051970 |
| 13 | 50.543155 | 63 | 244.939905 |
| 14 | 54.431090 | 64 | 248.827840 |
| 15 | 58.319025 | 65 | 252.715775 |
| 16 | 62.206960 | 66 | 256.603710 |
| 17 | 66.094895 | 67 | 260.491645 |
| 18 | 69.982830 | 68 | 264.379580 |
| 19 | 73.870765 | 69 | 268.267515 |
| 20 | 77.758700 | 70 | 272.155450 |
| 21 | 81.646635 | 71 | 276.043385 |
| 22 | 85.534570 | 72 | 279.931320 |
| 23 | 89.422505 | 73 | 283.819255 |
| 24 | 93.310440 | 74 | 287.707190 |
| 25 | 97.198375 | 75 | 291.595125 |
| 26 | 101.086310 | 76 | 295.483060 |
| 27 | 104.974245 | 77 | 299.370995 |
| 28 | 108.862180 | 78 | 303.258930 |
| 29 | 112.750115 | 79 | 307.146865 |
| 30 | 116.638050 | 80 | 311.034800 |
| 31 | 120.525985 | 81 | 314.922735 |
| 32 | 124.413920 | 82 | 318.810670 |
| 33 | 128.301855 | 83 | 322.698605 |
| 34 | 132.189790 | 84 | 326.586540 |
| 35 | 136.077725 | 85 | 330.474475 |
| 36 | 139.965660 | 86 | 334.362410 |
| 37 | 143.853595 | 87 | 338.250345 |
| 38 | 147.741530 | 88 | 342.138280 |
| 39 | 151.629465 | 89 | 346.026215 |
| 40 | 155.517400 | 90 | 349.914150 |
| 41 | 159.405335 | 91 | 353.802085 |
| 42 | 163.293270 | 92 | 357.690020 |
| 43 | 167.181205 | 93 | 361.577955 |
| 44 | 171.069140 | 94 | 365.465890 |
| 45 | 174.957075 | 95 | 369.353825 |
| 46 | 178.845010 | 96 | 373.241760 |
| 47 | 182.732945 | 97 | 377.129695 |
| 48 | 186.620880 | 98 | 381.017630 |
| 49 | 190.508815 | 99 | 384.905565 |
| 50 | 194.396750 | 100 | 388.793500 |

## WEIGHT
### Grams to Drams (apothecaries')

| g | dr ap | g | dr ap |
|---|---|---|---|
| 1 | .257206 | 51 | 13.117506 |
| 2 | .514412 | 52 | 13.374712 |
| 3 | .771618 | 53 | 13.631918 |
| 4 | 1.028824 | 54 | 13.889124 |
| 5 | 1.286030 | 55 | 14.146330 |
| 6 | 1.543236 | 56 | 14.403536 |
| 7 | 1.800442 | 57 | 14.660742 |
| 8 | 2.057648 | 58 | 14.917948 |
| 9 | 2.314854 | 59 | 15.175154 |
| 10 | 2.572060 | 60 | 15.432360 |
| 11 | 2.829266 | 61 | 15.689566 |
| 12 | 3.086472 | 62 | 15.946772 |
| 13 | 3.343678 | 63 | 16.203978 |
| 14 | 3.600884 | 64 | 16.461184 |
| 15 | 3.858090 | 65 | 16.718390 |
| 16 | 4.115296 | 66 | 16.975596 |
| 17 | 4.372502 | 67 | 17.232802 |
| 18 | 4.629708 | 68 | 17.490008 |
| 19 | 4.886914 | 69 | 17.747214 |
| 20 | 5.144120 | 70 | 18.004420 |
| 21 | 5.401326 | 71 | 18.261626 |
| 22 | 5.658532 | 72 | 18.518832 |
| 23 | 5.915738 | 73 | 18.776038 |
| 24 | 6.172944 | 74 | 19.033244 |
| 25 | 6.430150 | 75 | 19.290450 |
| 26 | 6.687356 | 76 | 19.547656 |
| 27 | 6.944562 | 77 | 19.804862 |
| 28 | 7.201768 | 78 | 20.062068 |
| 29 | 7.458974 | 79 | 20.319274 |
| 30 | 7.716180 | 80 | 20.576480 |
| 31 | 7.973386 | 81 | 20.833686 |
| 32 | 8.230592 | 82 | 21.090892 |
| 33 | 8.487798 | 83 | 21.348098 |
| 34 | 8.745004 | 84 | 21.605304 |
| 35 | 9.002210 | 85 | 21.862510 |
| 36 | 9.259416 | 86 | 22.119716 |
| 37 | 9.516622 | 87 | 22.376922 |
| 38 | 9.773828 | 88 | 22.634128 |
| 39 | 10.031034 | 89 | 22.891334 |
| 40 | 10.288240 | 90 | 23.148540 |
| 41 | 10.545446 | 91 | 23.405746 |
| 42 | 10.802652 | 92 | 23.662952 |
| 43 | 11.059858 | 93 | 23.920158 |
| 44 | 11.317064 | 94 | 24.177364 |
| 45 | 11.574270 | 95 | 24.434570 |
| 46 | 11.831476 | 96 | 24.691776 |
| 47 | 12.088682 | 97 | 24.948982 |
| 48 | 12.345888 | 98 | 25.206188 |
| 49 | 12.603094 | 99 | 25.463394 |
| 50 | 12.860300 | 100 | 25.720600 |

## WEIGHT
### Drams (avoirdupois) to Grams

| dr av | g | dr av | g |
|---|---|---|---|
| 1 | 1.771845 | 51 | 90.364095 |
| 2 | 3.543690 | 52 | 92.135940 |
| 3 | 5.315535 | 53 | 93.907785 |
| 4 | 7.087380 | 54 | 95.679630 |
| 5 | 8.859225 | 55 | 97.451475 |
| 6 | 10.631070 | 56 | 99.223320 |
| 7 | 12.402915 | 57 | 100.995165 |
| 8 | 14.174760 | 58 | 102.767010 |
| 9 | 15.946605 | 59 | 104.538855 |
| 10 | 17.718450 | 60 | 106.310700 |
| 11 | 19.490295 | 61 | 108.082545 |
| 12 | 21.262140 | 62 | 109.854390 |
| 13 | 23.033985 | 63 | 111.626235 |
| 14 | 24.805830 | 64 | 113.398080 |
| 15 | 26.577675 | 65 | 115.169925 |
| 16 | 28.349520 | 66 | 116.941770 |
| 17 | 30.121365 | 67 | 118.713615 |
| 18 | 31.893210 | 68 | 120.485460 |
| 19 | 33.665055 | 69 | 122.257305 |
| 20 | 35.436900 | 70 | 124.029150 |
| 21 | 37.208745 | 71 | 125.800995 |
| 22 | 38.980590 | 72 | 127.572840 |
| 23 | 40.752435 | 73 | 129.344685 |
| 24 | 42.524280 | 74 | 131.116530 |
| 25 | 44.296125 | 75 | 132.888375 |
| 26 | 46.067970 | 76 | 134.660220 |
| 27 | 47.839815 | 77 | 136.432065 |
| 28 | 49.611660 | 78 | 138.203910 |
| 29 | 51.383505 | 79 | 139.975755 |
| 30 | 53.155350 | 80 | 141.747600 |
| 31 | 54.927195 | 81 | 143.519445 |
| 32 | 56.699040 | 82 | 145.291290 |
| 33 | 58.470885 | 83 | 147.063135 |
| 34 | 60.242730 | 84 | 148.834980 |
| 35 | 62.014575 | 85 | 150.606825 |
| 36 | 63.786420 | 86 | 152.378670 |
| 37 | 65.558265 | 87 | 154.150515 |
| 38 | 67.330110 | 88 | 155.922360 |
| 39 | 69.101955 | 89 | 157.694205 |
| 40 | 70.873800 | 90 | 159.466050 |
| 41 | 72.645645 | 91 | 161.237895 |
| 42 | 74.417490 | 92 | 163.009740 |
| 43 | 76.189335 | 93 | 164.781585 |
| 44 | 77.961180 | 94 | 166.553430 |
| 45 | 79.733025 | 95 | 168.325275 |
| 46 | 81.504870 | 96 | 170.097120 |
| 47 | 83.276715 | 97 | 171.868965 |
| 48 | 85.048560 | 98 | 173.640810 |
| 49 | 86.820405 | 99 | 175.412655 |
| 50 | 88.592250 | 100 | 177.184500 |

## WEIGHT
### Grams to Drams (avoirdupois)

| g | dr av | g | dr av |
|---|---|---|---|
| 1 | .564383 | 51 | 28.783533 |
| 2 | 1.128766 | 52 | 29.347916 |
| 3 | 1.693149 | 53 | 29.912299 |
| 4 | 2.257532 | 54 | 30.476682 |
| 5 | 2.821915 | 55 | 31.041065 |
| 6 | 3.386298 | 56 | 31.605448 |
| 7 | 3.950681 | 57 | 32.169831 |
| 8 | 4.515064 | 58 | 32.734214 |
| 9 | 5.079447 | 59 | 33.298597 |
| 10 | 5.643830 | 60 | 33.862980 |
| 11 | 6.208213 | 61 | 34.427363 |
| 12 | 6.772596 | 62 | 34.991746 |
| 13 | 7.336979 | 63 | 35.556129 |
| 14 | 7.901362 | 64 | 36.120512 |
| 15 | 8.465745 | 65 | 36.684895 |
| 16 | 9.030128 | 66 | 37.249278 |
| 17 | 9.594511 | 67 | 37.813661 |
| 18 | 10.158894 | 68 | 38.378044 |
| 19 | 10.723277 | 69 | 38.942427 |
| 20 | 11.287660 | 70 | 39.506810 |
| 21 | 11.852043 | 71 | 40.071193 |
| 22 | 12.416426 | 72 | 40.635576 |
| 23 | 12.980809 | 73 | 41.199959 |
| 24 | 13.545192 | 74 | 41.764342 |
| 25 | 14.109575 | 75 | 42.328725 |
| 26 | 14.673958 | 76 | 42.893108 |
| 27 | 15.238341 | 77 | 43.457491 |
| 28 | 15.802724 | 78 | 44.021874 |
| 29 | 16.367107 | 79 | 44.586257 |
| 30 | 16.931490 | 80 | 45.150640 |
| 31 | 17.495873 | 81 | 45.715023 |
| 32 | 18.060256 | 82 | 46.279406 |
| 33 | 18.624639 | 83 | 46.843789 |
| 34 | 19.189022 | 84 | 47.408172 |
| 35 | 19.753405 | 85 | 47.972555 |
| 36 | 20.317788 | 86 | 48.536938 |
| 37 | 20.882171 | 87 | 49.101321 |
| 38 | 21.446554 | 88 | 49.665704 |
| 39 | 22.010937 | 89 | 50.230087 |
| 40 | 22.575320 | 90 | 50.794470 |
| 41 | 23.139703 | 91 | 51.358853 |
| 42 | 23.704086 | 92 | 51.923236 |
| 43 | 24.268469 | 93 | 52.487619 |
| 44 | 24.832852 | 94 | 53.052002 |
| 45 | 25.397235 | 95 | 53.616385 |
| 46 | 25.961618 | 96 | 54.180768 |
| 47 | 26.526001 | 97 | 54.745151 |
| 48 | 27.090384 | 98 | 55.309534 |
| 49 | 27.654767 | 99 | 55.873917 |
| 50 | 28.219150 | 100 | 56.438300 |

## WEIGHT
### Ounces (troy) to Grams

| oz t | g | oz t | g |
|---|---|---|---|
| 1 | 31.103480 | 51 | 1586.277480 |
| 2 | 62.206960 | 52 | 1617.380960 |
| 3 | 93.310440 | 53 | 1648.484440 |
| 4 | 124.413920 | 54 | 1679.587920 |
| 5 | 155.517400 | 55 | 1710.691400 |
| 6 | 186.620880 | 56 | 1741.794880 |
| 7 | 217.724360 | 57 | 1772.898360 |
| 8 | 248.827840 | 58 | 1804.001840 |
| 9 | 279.931320 | 59 | 1835.105320 |
| 10 | 311.034800 | 60 | 1866.208800 |
| 11 | 342.138280 | 61 | 1897.312280 |
| 12 | 373.241760 | 62 | 1928.415760 |
| 13 | 404.345240 | 63 | 1959.519240 |
| 14 | 435.448720 | 64 | 1990.622720 |
| 15 | 466.552200 | 65 | 2021.726200 |
| 16 | 497.655680 | 66 | 2052.829680 |
| 17 | 528.759160 | 67 | 2083.933160 |
| 18 | 559.862640 | 68 | 2115.036640 |
| 19 | 590.966120 | 69 | 2146.140120 |
| 20 | 622.069600 | 70 | 2177.243600 |
| 21 | 653.173080 | 71 | 2208.347080 |
| 22 | 684.276560 | 72 | 2239.450560 |
| 23 | 715.380040 | 73 | 2270.554040 |
| 24 | 746.483520 | 74 | 2301.657520 |
| 25 | 777.587000 | 75 | 2332.761000 |
| 26 | 808.690480 | 76 | 2363.864480 |
| 27 | 839.793960 | 77 | 2394.967960 |
| 28 | 870.897440 | 78 | 2426.071440 |
| 29 | 902.000920 | 79 | 2457.174920 |
| 30 | 933.104400 | 80 | 2488.278400 |
| 31 | 964.207880 | 81 | 2519.381880 |
| 32 | 995.311360 | 82 | 2550.485360 |
| 33 | 1026.414840 | 83 | 2581.588840 |
| 34 | 1057.518320 | 84 | 2612.692320 |
| 35 | 1088.621800 | 85 | 2643.795800 |
| 36 | 1119.725280 | 86 | 2674.899280 |
| 37 | 1150.828760 | 87 | 2706.002760 |
| 38 | 1181.932240 | 88 | 2737.106240 |
| 39 | 1213.035720 | 89 | 2768.209720 |
| 40 | 1244.139200 | 90 | 2799.313200 |
| 41 | 1275.242680 | 91 | 2830.416680 |
| 42 | 1306.346160 | 92 | 2861.520160 |
| 43 | 1337.449640 | 93 | 2892.623640 |
| 44 | 1368.553120 | 94 | 2923.727120 |
| 45 | 1399.656600 | 95 | 2954.830600 |
| 46 | 1430.760080 | 96 | 2985.934080 |
| 47 | 1461.863560 | 97 | 3017.037560 |
| 48 | 1492.967040 | 98 | 3048.141040 |
| 49 | 1524.070520 | 99 | 3079.244520 |
| 50 | 1555.174000 | 100 | 3110.348000 |

## WEIGHT
### Grams to Ounces (troy)

| g | oz t | g | oz t |
|---|---|---|---|
| 1 | .032151 | 51 | 1.639701 |
| 2 | .064302 | 52 | 1.671852 |
| 3 | .096453 | 53 | 1.704003 |
| 4 | .128604 | 54 | 1.736154 |
| 5 | .160755 | 55 | 1.768305 |
| 6 | .192906 | 56 | 1.800456 |
| 7 | .225057 | 57 | 1.832607 |
| 8 | .257208 | 58 | 1.864758 |
| 9 | .289359 | 59 | 1.896909 |
| 10 | .321510 | 60 | 1.929060 |
| 11 | .353661 | 61 | 1.961211 |
| 12 | .385812 | 62 | 1.993362 |
| 13 | .417963 | 63 | 2.025513 |
| 14 | .450114 | 64 | 2.057664 |
| 15 | .482265 | 65 | 2.089815 |
| 16 | .514416 | 66 | 2.121966 |
| 17 | .546567 | 67 | 2.154117 |
| 18 | .578718 | 68 | 2.186268 |
| 19 | .610869 | 69 | 2.218419 |
| 20 | .643020 | 70 | 2.250570 |
| 21 | .675171 | 71 | 2.282721 |
| 22 | .707322 | 72 | 2.314872 |
| 23 | .739473 | 73 | 2.347023 |
| 24 | .771624 | 74 | 2.379174 |
| 25 | .803775 | 75 | 2.411325 |
| 26 | .835926 | 76 | 2.443476 |
| 27 | .868077 | 77 | 2.475627 |
| 28 | .900228 | 78 | 2.507778 |
| 29 | .932379 | 79 | 2.539929 |
| 30 | .964530 | 80 | 2.572080 |
| 31 | .996681 | 81 | 2.604231 |
| 32 | 1.028832 | 82 | 2.636382 |
| 33 | 1.060983 | 83 | 2.668533 |
| 34 | 1.093134 | 84 | 2.700684 |
| 35 | 1.125285 | 85 | 2.732835 |
| 36 | 1.157436 | 86 | 2.764986 |
| 37 | 1.189587 | 87 | 2.797137 |
| 38 | 1.221738 | 88 | 2.829288 |
| 39 | 1.253889 | 89 | 2.861439 |
| 40 | 1.286040 | 90 | 2.893590 |
| 41 | 1.318191 | 91 | 2.925741 |
| 42 | 1.350342 | 92 | 2.957892 |
| 43 | 1.382493 | 93 | 2.990043 |
| 44 | 1.414644 | 94 | 3.022194 |
| 45 | 1.446795 | 95 | 3.054345 |
| 46 | 1.478946 | 96 | 3.086496 |
| 47 | 1.511097 | 97 | 3.118647 |
| 48 | 1.543248 | 98 | 3.150798 |
| 49 | 1.575399 | 99 | 3.182949 |
| 50 | 1.607550 | 100 | 3.215100 |

## WEIGHT
### Ounces (avoirdupois) to Grams

| oz av | g | oz av | g |
|---|---|---|---|
| 1 | 28.349520 | 51 | 1445.825520 |
| 2 | 56.699040 | 52 | 1474.175040 |
| 3 | 85.048560 | 53 | 1502.524560 |
| 4 | 113.398080 | 54 | 1530.874080 |
| 5 | 141.747600 | 55 | 1559.223600 |
| 6 | 170.097120 | 56 | 1587.573120 |
| 7 | 198.446640 | 57 | 1615.922640 |
| 8 | 226.796160 | 58 | 1644.272160 |
| 9 | 255.145680 | 59 | 1672.621680 |
| 10 | 283.495200 | 60 | 1700.971200 |
| 11 | 311.844720 | 61 | 1729.320720 |
| 12 | 340.194240 | 62 | 1757.670240 |
| 13 | 368.543760 | 63 | 1786.019760 |
| 14 | 396.893280 | 64 | 1814.369280 |
| 15 | 425.242800 | 65 | 1842.718800 |
| 16 | 453.592320 | 66 | 1871.068320 |
| 17 | 481.941840 | 67 | 1899.417840 |
| 18 | 510.291360 | 68 | 1927.767360 |
| 19 | 538.640880 | 69 | 1956.116880 |
| 20 | 566.990400 | 70 | 1984.466400 |
| 21 | 595.339920 | 71 | 2012.815920 |
| 22 | 623.689440 | 72 | 2041.165440 |
| 23 | 652.038960 | 73 | 2069.514960 |
| 24 | 680.388480 | 74 | 2097.864480 |
| 25 | 708.738000 | 75 | 2126.214000 |
| 26 | 737.087520 | 76 | 2154.563520 |
| 27 | 765.437040 | 77 | 2182.913040 |
| 28 | 793.786560 | 78 | 2211.262560 |
| 29 | 822.136080 | 79 | 2239.612080 |
| 30 | 850.485600 | 80 | 2267.961600 |
| 31 | 878.835120 | 81 | 2296.311120 |
| 32 | 907.184640 | 82 | 2324.660640 |
| 33 | 935.534160 | 83 | 2353.010160 |
| 34 | 963.883680 | 84 | 2381.359680 |
| 35 | 992.233200 | 85 | 2409.709200 |
| 36 | 1020.582720 | 86 | 2438.058720 |
| 37 | 1048.932240 | 87 | 2466.408240 |
| 38 | 1077.281760 | 88 | 2494.757760 |
| 39 | 1105.631280 | 89 | 2523.107280 |
| 40 | 1133.980800 | 90 | 2551.456800 |
| 41 | 1162.330320 | 91 | 2579.806320 |
| 42 | 1190.679840 | 92 | 2608.155840 |
| 43 | 1219.029360 | 93 | 2636.505360 |
| 44 | 1247.378880 | 94 | 2664.854880 |
| 45 | 1275.728400 | 95 | 2693.204400 |
| 46 | 1304.077920 | 96 | 2721.553920 |
| 47 | 1332.427440 | 97 | 2749.903440 |
| 48 | 1360.776960 | 98 | 2778.252960 |
| 49 | 1389.126480 | 99 | 2806.602480 |
| 50 | 1417.476000 | 100 | 2834.952000 |

## WEIGHT
### Grams to Ounces (avoirdupois)

| g | oz av | g | oz av |
|---|---|---|---|
| 1 | .035274 | 51 | 1.798974 |
| 2 | .070548 | 52 | 1.834248 |
| 3 | .105822 | 53 | 1.869522 |
| 4 | .141096 | 54 | 1.904796 |
| 5 | .176370 | 55 | 1.940070 |
| 6 | .211644 | 56 | 1.975344 |
| 7 | .246918 | 57 | 2.010618 |
| 8 | .282192 | 58 | 2.045892 |
| 9 | .317466 | 59 | 2.081166 |
| 10 | .352740 | 60 | 2.116440 |
| 11 | .388014 | 61 | 2.151714 |
| 12 | .423288 | 62 | 2.186988 |
| 13 | .458562 | 63 | 2.222262 |
| 14 | .493836 | 64 | 2.257536 |
| 15 | .529110 | 65 | 2.292810 |
| 16 | .564384 | 66 | 2.328084 |
| 17 | .599658 | 67 | 2.363358 |
| 18 | .634932 | 68 | 2.398632 |
| 19 | .670206 | 69 | 2.433906 |
| 20 | .705480 | 70 | 2.469180 |
| 21 | .740754 | 71 | 2.504454 |
| 22 | .776028 | 72 | 2.539728 |
| 23 | .811302 | 73 | 2.575002 |
| 24 | .846576 | 74 | 2.610276 |
| 25 | .881850 | 75 | 2.645550 |
| 26 | .917124 | 76 | 2.680824 |
| 27 | .952398 | 77 | 2.716098 |
| 28 | .987672 | 78 | 2.751372 |
| 29 | 1.022946 | 79 | 2.786646 |
| 30 | 1.058220 | 80 | 2.821920 |
| 31 | 1.093494 | 81 | 2.857194 |
| 32 | 1.128768 | 82 | 2.892468 |
| 33 | 1.164042 | 83 | 2.927742 |
| 34 | 1.199316 | 84 | 2.963016 |
| 35 | 1.234590 | 85 | 2.998290 |
| 36 | 1.269864 | 86 | 3.033564 |
| 37 | 1.305138 | 87 | 3.068838 |
| 38 | 1.340412 | 88 | 3.104112 |
| 39 | 1.375686 | 89 | 3.139386 |
| 40 | 1.410960 | 90 | 3.174660 |
| 41 | 1.446234 | 91 | 3.209934 |
| 42 | 1.481508 | 92 | 3.245208 |
| 43 | 1.516782 | 93 | 3.280482 |
| 44 | 1.552056 | 94 | 3.315756 |
| 45 | 1.587330 | 95 | 3.351030 |
| 46 | 1.622604 | 96 | 3.386304 |
| 47 | 1.657878 | 97 | 3.421578 |
| 48 | 1.693152 | 98 | 3.456852 |
| 49 | 1.728426 | 99 | 3.492126 |
| 50 | 1.763700 | 100 | 3.527400 |

## WEIGHT
### Pounds (troy) to Kilograms

| lb t | kg | lb t | kg |
|---|---|---|---|
| 1 | .373242 | 51 | 19.035342 |
| 2 | .746484 | 52 | 19.408584 |
| 3 | 1.119726 | 53 | 19.781826 |
| 4 | 1.492968 | 54 | 20.155068 |
| 5 | 1.866210 | 55 | 20.528310 |
| 6 | 2.239452 | 56 | 20.901552 |
| 7 | 2.612694 | 57 | 21.274794 |
| 8 | 2.985936 | 58 | 21.648036 |
| 9 | 3.359178 | 59 | 22.021278 |
| 10 | 3.732420 | 60 | 22.394520 |
| 11 | 4.105662 | 61 | 22.767762 |
| 12 | 4.478904 | 62 | 23.141004 |
| 13 | 4.852146 | 63 | 23.514246 |
| 14 | 5.225388 | 64 | 23.887488 |
| 15 | 5.598630 | 65 | 24.260730 |
| 16 | 5.971872 | 66 | 24.633972 |
| 17 | 6.345114 | 67 | 25.007214 |
| 18 | 6.718356 | 68 | 25.380456 |
| 19 | 7.091598 | 69 | 25.753698 |
| 20 | 7.464840 | 70 | 26.126940 |
| 21 | 7.838082 | 71 | 26.500182 |
| 22 | 8.211324 | 72 | 26.873424 |
| 23 | 8.584566 | 73 | 27.246666 |
| 24 | 8.957808 | 74 | 27.619908 |
| 25 | 9.331050 | 75 | 27.993150 |
| 26 | 9.704292 | 76 | 28.366392 |
| 27 | 10.077534 | 77 | 28.739634 |
| 28 | 10.450776 | 78 | 29.112876 |
| 29 | 10.824018 | 79 | 29.486118 |
| 30 | 11.197260 | 80 | 29.859360 |
| 31 | 11.570502 | 81 | 30.232602 |
| 32 | 11.943744 | 82 | 30.605844 |
| 33 | 12.316986 | 83 | 30.979086 |
| 34 | 12.690228 | 84 | 31.352328 |
| 35 | 13.063470 | 85 | 31.725570 |
| 36 | 13.436712 | 86 | 32.098812 |
| 37 | 13.809954 | 87 | 32.472054 |
| 38 | 14.183196 | 88 | 32.845296 |
| 39 | 14.556438 | 89 | 33.218538 |
| 40 | 14.929680 | 90 | 33.591780 |
| 41 | 15.302922 | 91 | 33.965022 |
| 42 | 15.676164 | 92 | 34.338264 |
| 43 | 16.049406 | 93 | 34.711506 |
| 44 | 16.422648 | 94 | 35.084748 |
| 45 | 16.795890 | 95 | 35.457990 |
| 46 | 17.169132 | 96 | 35.831232 |
| 47 | 17.542374 | 97 | 36.204474 |
| 48 | 17.915616 | 98 | 36.577716 |
| 49 | 18.288858 | 99 | 36.950958 |
| 50 | 18.662100 | 100 | 37.324200 |

## WEIGHT
### Kilograms to Pounds (troy)

| kg | lb t | kg | lb t |
|---|---|---|---|
| 1 | 2.679227 | 51 | 136.640577 |
| 2 | 5.358454 | 52 | 139.319804 |
| 3 | 8.037681 | 53 | 141.999031 |
| 4 | 10.716908 | 54 | 144.678258 |
| 5 | 13.396135 | 55 | 147.357485 |
| 6 | 16.075362 | 56 | 150.036712 |
| 7 | 18.754589 | 57 | 152.715939 |
| 8 | 21.433816 | 58 | 155.395166 |
| 9 | 24.113043 | 59 | 158.074393 |
| 10 | 26.792270 | 60 | 160.753620 |
| 11 | 29.471497 | 61 | 163.432847 |
| 12 | 32.150724 | 62 | 166.112074 |
| 13 | 34.829951 | 63 | 168.791301 |
| 14 | 37.509178 | 64 | 171.470528 |
| 15 | 40.188405 | 65 | 174.149755 |
| 16 | 42.867632 | 66 | 176.828982 |
| 17 | 45.546859 | 67 | 179.508209 |
| 18 | 48.226086 | 68 | 182.187436 |
| 19 | 50.905313 | 69 | 184.866663 |
| 20 | 53.584540 | 70 | 187.545890 |
| 21 | 56.263767 | 71 | 190.225117 |
| 22 | 58.942994 | 72 | 192.904344 |
| 23 | 61.622221 | 73 | 195.583571 |
| 24 | 64.301448 | 74 | 198.262798 |
| 25 | 66.980675 | 75 | 200.942025 |
| 26 | 69.659902 | 76 | 203.621252 |
| 27 | 72.339129 | 77 | 206.300479 |
| 28 | 75.018356 | 78 | 208.979706 |
| 29 | 77.697583 | 79 | 211.658933 |
| 30 | 80.376810 | 80 | 214.338160 |
| 31 | 83.056037 | 81 | 217.017387 |
| 32 | 85.735264 | 82 | 219.696614 |
| 33 | 88.414491 | 83 | 222.375841 |
| 34 | 91.093718 | 84 | 225.055068 |
| 35 | 93.772945 | 85 | 227.734295 |
| 36 | 96.452172 | 86 | 230.413522 |
| 37 | 99.131399 | 87 | 233.092749 |
| 38 | 101.810626 | 88 | 235.771976 |
| 39 | 104.489853 | 89 | 238.451203 |
| 40 | 107.169080 | 90 | 241.130430 |
| 41 | 109.848307 | 91 | 243.809657 |
| 42 | 112.527534 | 92 | 246.488884 |
| 43 | 115.206761 | 93 | 249.168111 |
| 44 | 117.885988 | 94 | 251.847338 |
| 45 | 120.565215 | 95 | 254.526565 |
| 46 | 123.244442 | 96 | 257.205792 |
| 47 | 125.923669 | 97 | 259.885019 |
| 48 | 128.602896 | 98 | 262.564246 |
| 49 | 131.282123 | 99 | 265.243473 |
| 50 | 133.961350 | 100 | 267.922700 |

## WEIGHT
### Pounds (avoirdupois) to Kilograms

| lb | kg | lb | kg |
|---|---|---|---|
| 1 | .453592 | 51 | 23.133192 |
| 2 | .907184 | 52 | 23.586784 |
| 3 | 1.360776 | 53 | 24.040376 |
| 4 | 1.814368 | 54 | 24.493968 |
| 5 | 2.267960 | 55 | 24.947560 |
| 6 | 2.721552 | 56 | 25.401152 |
| 7 | 3.175144 | 57 | 25.854744 |
| 8 | 3.628736 | 58 | 26.308336 |
| 9 | 4.082328 | 59 | 26.761928 |
| 10 | 4.535920 | 60 | 27.215520 |
| 11 | 4.989512 | 61 | 27.669112 |
| 12 | 5.443104 | 62 | 28.122704 |
| 13 | 5.896696 | 63 | 28.576296 |
| 14 | 6.350288 | 64 | 29.029888 |
| 15 | 6.803880 | 65 | 29.483480 |
| 16 | 7.257472 | 66 | 29.937072 |
| 17 | 7.711064 | 67 | 30.390664 |
| 18 | 8.164656 | 68 | 30.844256 |
| 19 | 8.618248 | 69 | 31.297848 |
| 20 | 9.071840 | 70 | 31.751440 |
| 21 | 9.525432 | 71 | 32.205032 |
| 22 | 9.979024 | 72 | 32.658624 |
| 23 | 10.432616 | 73 | 33.112216 |
| 24 | 10.886208 | 74 | 33.565808 |
| 25 | 11.339800 | 75 | 34.019400 |
| 26 | 11.793392 | 76 | 34.472992 |
| 27 | 12.246984 | 77 | 34.926584 |
| 28 | 12.700576 | 78 | 35.380176 |
| 29 | 13.154168 | 79 | 35.833768 |
| 30 | 13.607760 | 80 | 36.287360 |
| 31 | 14.061352 | 81 | 36.740952 |
| 32 | 14.514944 | 82 | 37.194544 |
| 33 | 14.968536 | 83 | 37.648136 |
| 34 | 15.422128 | 84 | 38.101728 |
| 35 | 15.875720 | 85 | 38.555320 |
| 36 | 16.329312 | 86 | 39.008912 |
| 37 | 16.782904 | 87 | 39.462504 |
| 38 | 17.236496 | 88 | 39.916096 |
| 39 | 17.690088 | 89 | 40.369688 |
| 40 | 18.143680 | 90 | 40.823280 |
| 41 | 18.597272 | 91 | 41.276872 |
| 42 | 19.050864 | 92 | 41.730464 |
| 43 | 19.504456 | 93 | 42.184056 |
| 44 | 19.958048 | 94 | 42.637648 |
| 45 | 20.411640 | 95 | 43.091240 |
| 46 | 20.865232 | 96 | 43.544832 |
| 47 | 21.318824 | 97 | 43.998424 |
| 48 | 21.772416 | 98 | 44.452016 |
| 49 | 22.226008 | 99 | 44.905608 |
| 50 | 22.679600 | 100 | 45.359200 |

## WEIGHT
### Kilograms to Pounds (avoirdupois)

| kg | lb | kg | lb |
|---|---|---|---|
| 1 | 2.204622 | 51 | 112.435722 |
| 2 | 4.409244 | 52 | 114.640344 |
| 3 | 6.613866 | 53 | 116.844966 |
| 4 | 8.818488 | 54 | 119.049588 |
| 5 | 11.023110 | 55 | 121.254210 |
| 6 | 13.227732 | 56 | 123.458832 |
| 7 | 15.432354 | 57 | 125.663454 |
| 8 | 17.636976 | 58 | 127.868076 |
| 9 | 19.841598 | 59 | 130.072698 |
| 10 | 22.046220 | 60 | 132.277320 |
| 11 | 24.250842 | 61 | 134.481942 |
| 12 | 26.455464 | 62 | 136.686564 |
| 13 | 28.660086 | 63 | 138.891186 |
| 14 | 30.864708 | 64 | 141.095808 |
| 15 | 33.069330 | 65 | 143.300430 |
| 16 | 35.273952 | 66 | 145.505052 |
| 17 | 37.478574 | 67 | 147.709674 |
| 18 | 39.683196 | 68 | 149.914296 |
| 19 | 41.887818 | 69 | 152.118918 |
| 20 | 44.092440 | 70 | 154.323540 |
| 21 | 46.297062 | 71 | 156.528162 |
| 22 | 48.501684 | 72 | 158.732784 |
| 23 | 50.706306 | 73 | 160.937406 |
| 24 | 52.910928 | 74 | 163.142028 |
| 25 | 55.115550 | 75 | 165.346650 |
| 26 | 57.320172 | 76 | 167.551272 |
| 27 | 59.524794 | 77 | 169.755894 |
| 28 | 61.729416 | 78 | 171.960516 |
| 29 | 63.934038 | 79 | 174.165138 |
| 30 | 66.138660 | 80 | 176.369760 |
| 31 | 68.343282 | 81 | 178.574382 |
| 32 | 70.547904 | 82 | 180.779004 |
| 33 | 72.752526 | 83 | 182.983626 |
| 34 | 74.957148 | 84 | 185.188248 |
| 35 | 77.161770 | 85 | 187.392870 |
| 36 | 79.366392 | 86 | 189.597492 |
| 37 | 81.571014 | 87 | 191.802114 |
| 38 | 83.775636 | 88 | 194.006736 |
| 39 | 85.980258 | 89 | 196.211358 |
| 40 | 88.184880 | 90 | 198.415980 |
| 41 | 90.389502 | 91 | 200.620602 |
| 42 | 92.594124 | 92 | 202.825224 |
| 43 | 94.798746 | 93 | 205.029846 |
| 44 | 97.003368 | 94 | 207.234468 |
| 45 | 99.207990 | 95 | 209.439090 |
| 46 | 101.412612 | 96 | 211.643712 |
| 47 | 103.617234 | 97 | 213.848334 |
| 48 | 105.821856 | 98 | 216.052956 |
| 49 | 108.026478 | 99 | 218.257578 |
| 50 | 110.231100 | 100 | 220.462200 |

## WEIGHT
### Hundredweight (short) to Kilograms

| cwt | kg | cwt | kg |
|---|---|---|---|
| 1 | 45.359230 | 51 | 2313.320730 |
| 2 | 90.718460 | 52 | 2358.679960 |
| 3 | 136.077690 | 53 | 2404.039190 |
| 4 | 181.436920 | 54 | 2449.398420 |
| 5 | 226.796150 | 55 | 2494.757650 |
| 6 | 272.155380 | 56 | 2540.116880 |
| 7 | 317.514610 | 57 | 2585.476110 |
| 8 | 362.873840 | 58 | 2630.835340 |
| 9 | 408.233070 | 59 | 2676.194570 |
| 10 | 453.592300 | 60 | 2721.553800 |
| 11 | 498.951530 | 61 | 2766.913030 |
| 12 | 544.310760 | 62 | 2812.272260 |
| 13 | 589.669990 | 63 | 2857.631490 |
| 14 | 635.029220 | 64 | 2902.990720 |
| 15 | 680.388450 | 65 | 2948.349950 |
| 16 | 725.747680 | 66 | 2993.709180 |
| 17 | 771.106910 | 67 | 3039.068410 |
| 18 | 816.466140 | 68 | 3084.427640 |
| 19 | 861.825370 | 69 | 3129.786870 |
| 20 | 907.184600 | 70 | 3175.146100 |
| 21 | 952.543830 | 71 | 3220.505330 |
| 22 | 997.903060 | 72 | 3265.864560 |
| 23 | 1043.262290 | 73 | 3311.223790 |
| 24 | 1088.621520 | 74 | 3356.583020 |
| 25 | 1133.980750 | 75 | 3401.942250 |
| 26 | 1179.339980 | 76 | 3447.301480 |
| 27 | 1224.699210 | 77 | 3492.660710 |
| 28 | 1270.058440 | 78 | 3538.019940 |
| 29 | 1315.417670 | 79 | 3583.379170 |
| 30 | 1360.776900 | 80 | 3628.738400 |
| 31 | 1406.136130 | 81 | 3674.097630 |
| 32 | 1451.495360 | 82 | 3719.456860 |
| 33 | 1496.854590 | 83 | 3764.816090 |
| 34 | 1542.213820 | 84 | 3810.175320 |
| 35 | 1587.573050 | 85 | 3855.534550 |
| 36 | 1632.932280 | 86 | 3900.893780 |
| 37 | 1678.291510 | 87 | 3946.253010 |
| 38 | 1723.650740 | 88 | 3991.612240 |
| 39 | 1769.009970 | 89 | 4036.971470 |
| 40 | 1814.369200 | 90 | 4082.330700 |
| 41 | 1859.728430 | 91 | 4127.689930 |
| 42 | 1905.087660 | 92 | 4173.049160 |
| 43 | 1950.446890 | 93 | 4218.408390 |
| 44 | 1995.806120 | 94 | 4263.767620 |
| 45 | 2041.165350 | 95 | 4309.126850 |
| 46 | 2086.524580 | 96 | 4354.486080 |
| 47 | 2131.883810 | 97 | 4399.845310 |
| 48 | 2177.243040 | 98 | 4445.204540 |
| 49 | 2222.602270 | 99 | 4490.563770 |
| 50 | 2267.961500 | 100 | 4535.923000 |

## WEIGHT
### Kilograms to Hundredweight (short)

| kg | cwt | kg | cwt |
|---|---|---|---|
| 1 | .022046 | 51 | 1.124346 |
| 2 | .044092 | 52 | 1.146392 |
| 3 | .066138 | 53 | 1.168438 |
| 4 | .088184 | 54 | 1.190484 |
| 5 | .110230 | 55 | 1.212530 |
| 6 | .132276 | 56 | 1.234576 |
| 7 | .154322 | 57 | 1.256622 |
| 8 | .176368 | 58 | 1.278668 |
| 9 | .198414 | 59 | 1.300714 |
| 10 | .220460 | 60 | 1.322760 |
| 11 | .242506 | 61 | 1.344806 |
| 12 | .264552 | 62 | 1.366852 |
| 13 | .286598 | 63 | 1.388898 |
| 14 | .308644 | 64 | 1.410944 |
| 15 | .330690 | 65 | 1.432990 |
| 16 | .352736 | 66 | 1.455036 |
| 17 | .374782 | 67 | 1.477082 |
| 18 | .396828 | 68 | 1.499128 |
| 19 | .418874 | 69 | 1.521174 |
| 20 | .440920 | 70 | 1.543220 |
| 21 | .462966 | 71 | 1.565266 |
| 22 | .485012 | 72 | 1.587312 |
| 23 | .507058 | 73 | 1.609358 |
| 24 | .529104 | 74 | 1.631404 |
| 25 | .551150 | 75 | 1.653450 |
| 26 | .573196 | 76 | 1.675496 |
| 27 | .595242 | 77 | 1.697542 |
| 28 | .617288 | 78 | 1.719588 |
| 29 | .639334 | 79 | 1.741634 |
| 30 | .661380 | 80 | 1.763680 |
| 31 | .683426 | 81 | 1.785726 |
| 32 | .705472 | 82 | 1.807772 |
| 33 | .727518 | 83 | 1.829818 |
| 34 | .749564 | 84 | 1.851864 |
| 35 | .771610 | 85 | 1.873910 |
| 36 | .793656 | 86 | 1.895956 |
| 37 | .815702 | 87 | 1.918002 |
| 38 | .837748 | 88 | 1.940048 |
| 39 | .859794 | 89 | 1.962094 |
| 40 | .881840 | 90 | 1.984140 |
| 41 | .903886 | 91 | 2.006186 |
| 42 | .925932 | 92 | 2.028232 |
| 43 | .947978 | 93 | 2.050278 |
| 44 | .970024 | 94 | 2.072324 |
| 45 | .992070 | 95 | 2.094370 |
| 46 | 1.014116 | 96 | 2.116416 |
| 47 | 1.036162 | 97 | 2.138462 |
| 48 | 1.058208 | 98 | 2.160508 |
| 49 | 1.080254 | 99 | 2.182554 |
| 50 | 1.102300 | 100 | 2.204600 |

## WEIGHT
### Hundredweight (long) to Kilograms

| cwt | kg | cwt | kg |
|---|---|---|---|
| 1 | 50.802350 | 51 | 2590.919850 |
| 2 | 101.604700 | 52 | 2641.722200 |
| 3 | 152.407050 | 53 | 2692.524550 |
| 4 | 203.209400 | 54 | 2743.326900 |
| 5 | 254.011750 | 55 | 2794.129250 |
| 6 | 304.814100 | 56 | 2844.931600 |
| 7 | 355.616450 | 57 | 2895.733950 |
| 8 | 406.418800 | 58 | 2946.536300 |
| 9 | 457.221150 | 59 | 2997.338650 |
| 10 | 508.023500 | 60 | 3048.141000 |
| 11 | 558.825850 | 61 | 3098.943350 |
| 12 | 609.628200 | 62 | 3149.745700 |
| 13 | 660.430550 | 63 | 3200.548050 |
| 14 | 711.232900 | 64 | 3251.350400 |
| 15 | 762.035250 | 65 | 3302.152750 |
| 16 | 812.837600 | 66 | 3352.955100 |
| 17 | 863.639950 | 67 | 3403.757450 |
| 18 | 914.442300 | 68 | 3454.559800 |
| 19 | 965.244650 | 69 | 3505.362150 |
| 20 | 1016.047000 | 70 | 3556.164500 |
| 21 | 1066.849350 | 71 | 3606.966850 |
| 22 | 1117.651700 | 72 | 3657.769200 |
| 23 | 1168.454050 | 73 | 3708.571550 |
| 24 | 1219.256400 | 74 | 3759.373900 |
| 25 | 1270.058750 | 75 | 3810.176250 |
| 26 | 1320.861100 | 76 | 3860.978600 |
| 27 | 1371.663450 | 77 | 3911.780950 |
| 28 | 1422.465800 | 78 | 3962.583300 |
| 29 | 1473.268150 | 79 | 4013.385650 |
| 30 | 1524.070500 | 80 | 4064.188000 |
| 31 | 1574.872850 | 81 | 4114.990350 |
| 32 | 1625.675200 | 82 | 4165.792700 |
| 33 | 1676.477550 | 83 | 4216.595050 |
| 34 | 1727.279900 | 84 | 4267.397400 |
| 35 | 1778.082250 | 85 | 4318.199750 |
| 36 | 1828.884600 | 86 | 4369.002100 |
| 37 | 1879.686950 | 87 | 4419.804450 |
| 38 | 1930.489300 | 88 | 4470.606800 |
| 39 | 1981.291650 | 89 | 4521.409150 |
| 40 | 2032.094000 | 90 | 4572.211500 |
| 41 | 2082.896350 | 91 | 4623.013850 |
| 42 | 2133.698700 | 92 | 4673.816200 |
| 43 | 2184.501050 | 93 | 4724.618550 |
| 44 | 2235.303400 | 94 | 4775.420900 |
| 45 | 2286.105750 | 95 | 4826.223250 |
| 46 | 2336.908100 | 96 | 4877.025600 |
| 47 | 2387.710450 | 97 | 4927.827950 |
| 48 | 2438.512800 | 98 | 4978.630300 |
| 49 | 2489.315150 | 99 | 5029.432650 |
| 50 | 2540.117500 | 100 | 5080.235000 |

## WEIGHT
### Kilograms to Hundredweight (long)

| kg | cwt | kg | cwt |
|---|---|---|---|
| 1 | .019684 | 51 | 1.003884 |
| 2 | .039368 | 52 | 1.023568 |
| 3 | .059052 | 53 | 1.043252 |
| 4 | .078736 | 54 | 1.062936 |
| 5 | .098420 | 55 | 1.082620 |
| 6 | .118104 | 56 | 1.102304 |
| 7 | .137788 | 57 | 1.121988 |
| 8 | .157472 | 58 | 1.141672 |
| 9 | .177156 | 59 | 1.161356 |
| 10 | .196840 | 60 | 1.181040 |
| 11 | .216524 | 61 | 1.200724 |
| 12 | .236208 | 62 | 1.220408 |
| 13 | .255892 | 63 | 1.240092 |
| 14 | .275576 | 64 | 1.259776 |
| 15 | .295260 | 65 | 1.279460 |
| 16 | .314944 | 66 | 1.299144 |
| 17 | .334628 | 67 | 1.318828 |
| 18 | .354312 | 68 | 1.338512 |
| 19 | .373996 | 69 | 1.358196 |
| 20 | .393680 | 70 | 1.377880 |
| 21 | .413364 | 71 | 1.397564 |
| 22 | .433048 | 72 | 1.417248 |
| 23 | .452732 | 73 | 1.436932 |
| 24 | .472416 | 74 | 1.456616 |
| 25 | .492100 | 75 | 1.476300 |
| 26 | .511784 | 76 | 1.495984 |
| 27 | .531468 | 77 | 1.515668 |
| 28 | .551152 | 78 | 1.535352 |
| 29 | .570836 | 79 | 1.555036 |
| 30 | .590520 | 80 | 1.574720 |
| 31 | .610204 | 81 | 1.594404 |
| 32 | .629888 | 82 | 1.614088 |
| 33 | .649572 | 83 | 1.633772 |
| 34 | .669256 | 84 | 1.653456 |
| 35 | .688940 | 85 | 1.673140 |
| 36 | .708624 | 86 | 1.692824 |
| 37 | .728308 | 87 | 1.712508 |
| 38 | .747992 | 88 | 1.732192 |
| 39 | .767676 | 89 | 1.751876 |
| 40 | .787360 | 90 | 1.771560 |
| 41 | .807044 | 91 | 1.791244 |
| 42 | .826728 | 92 | 1.810928 |
| 43 | .846412 | 93 | 1.830612 |
| 44 | .866096 | 94 | 1.850296 |
| 45 | .885780 | 95 | 1.869980 |
| 46 | .905464 | 96 | 1.889664 |
| 47 | .925148 | 97 | 1.909348 |
| 48 | .944832 | 98 | 1.929032 |
| 49 | .964516 | 99 | 1.948716 |
| 50 | .984200 | 100 | 1.968400 |

## WEIGHT
### Short Ton to Metric Ton

| s ton | mt ton | s ton | mt ton |
|---|---|---|---|
| 1 | .907185 | 51 | 46.266435 |
| 2 | 1.814370 | 52 | 47.173620 |
| 3 | 2.721555 | 53 | 48.080805 |
| 4 | 3.628740 | 54 | 48.987990 |
| 5 | 4.535925 | 55 | 49.895175 |
| 6 | 5.443110 | 56 | 50.802360 |
| 7 | 6.350295 | 57 | 51.709545 |
| 8 | 7.257480 | 58 | 52.616730 |
| 9 | 8.164665 | 59 | 53.523915 |
| 10 | 9.071850 | 60 | 54.431100 |
| 11 | 9.979035 | 61 | 55.338285 |
| 12 | 10.886220 | 62 | 56.245470 |
| 13 | 11.793405 | 63 | 57.152655 |
| 14 | 12.700590 | 64 | 58.059840 |
| 15 | 13.607775 | 65 | 58.967025 |
| 16 | 14.514960 | 66 | 59.874210 |
| 17 | 15.422145 | 67 | 60.781395 |
| 18 | 16.329330 | 68 | 61.688580 |
| 19 | 17.236515 | 69 | 62.595765 |
| 20 | 18.143700 | 70 | 63.502950 |
| 21 | 19.050885 | 71 | 64.410135 |
| 22 | 19.958070 | 72 | 65.317320 |
| 23 | 20.865255 | 73 | 66.224505 |
| 24 | 21.772440 | 74 | 67.131690 |
| 25 | 22.679625 | 75 | 68.038875 |
| 26 | 23.586810 | 76 | 68.946060 |
| 27 | 24.493995 | 77 | 69.853245 |
| 28 | 25.401180 | 78 | 70.760430 |
| 29 | 26.308365 | 79 | 71.667615 |
| 30 | 27.215550 | 80 | 72.574800 |
| 31 | 28.122735 | 81 | 73.481985 |
| 32 | 29.029920 | 82 | 74.389170 |
| 33 | 29.937105 | 83 | 75.296355 |
| 34 | 30.844290 | 84 | 76.203540 |
| 35 | 31.751475 | 85 | 77.110725 |
| 36 | 32.658660 | 86 | 78.017910 |
| 37 | 33.565845 | 87 | 78.925095 |
| 38 | 34.473030 | 88 | 79.832280 |
| 39 | 35.380215 | 89 | 80.739465 |
| 40 | 36.287400 | 90 | 81.646650 |
| 41 | 37.194585 | 91 | 82.553835 |
| 42 | 38.101770 | 92 | 83.461020 |
| 43 | 39.008955 | 93 | 84.368205 |
| 44 | 39.916140 | 94 | 85.275390 |
| 45 | 40.823325 | 95 | 86.182575 |
| 46 | 41.730510 | 96 | 87.089760 |
| 47 | 42.637695 | 97 | 87.996945 |
| 48 | 43.544880 | 98 | 88.904130 |
| 49 | 44.452065 | 99 | 89.811315 |
| 50 | 45.359250 | 100 | 90.718500 |

## WEIGHT
### Metric Ton to Short Ton

| mt ton | s ton | mt ton | s ton |
|---|---|---|---|
| 1 | 1.102311 | 51 | 56.217861 |
| 2 | 2.204622 | 52 | 57.320172 |
| 3 | 3.306933 | 53 | 58.422483 |
| 4 | 4.409244 | 54 | 59.524794 |
| 5 | 5.511555 | 55 | 60.627105 |
| 6 | 6.613866 | 56 | 61.729416 |
| 7 | 7.716177 | 57 | 62.831727 |
| 8 | 8.818488 | 58 | 63.934038 |
| 9 | 9.920799 | 59 | 65.036349 |
| 10 | 11.023110 | 60 | 66.138660 |
| 11 | 12.125421 | 61 | 67.240971 |
| 12 | 13.227732 | 62 | 68.343282 |
| 13 | 14.330043 | 63 | 69.445593 |
| 14 | 15.432354 | 64 | 70.547904 |
| 15 | 16.534665 | 65 | 71.650215 |
| 16 | 17.636976 | 66 | 72.752526 |
| 17 | 18.739287 | 67 | 73.854837 |
| 18 | 19.841598 | 68 | 74.957148 |
| 19 | 20.943909 | 69 | 76.059459 |
| 20 | 22.046220 | 70 | 77.161770 |
| 21 | 23.148531 | 71 | 78.264081 |
| 22 | 24.250842 | 72 | 79.366392 |
| 23 | 25.353153 | 73 | 80.468703 |
| 24 | 26.455464 | 74 | 81.571014 |
| 25 | 27.557775 | 75 | 82.673325 |
| 26 | 28.660086 | 76 | 83.775636 |
| 27 | 29.762397 | 77 | 84.877947 |
| 28 | 30.864708 | 78 | 85.980258 |
| 29 | 31.967019 | 79 | 87.082569 |
| 30 | 33.069330 | 80 | 88.184880 |
| 31 | 34.171641 | 81 | 89.287191 |
| 32 | 35.273952 | 82 | 90.389502 |
| 33 | 36.376263 | 83 | 91.491813 |
| 34 | 37.478574 | 84 | 92.594124 |
| 35 | 38.580885 | 85 | 93.696435 |
| 36 | 39.683196 | 86 | 94.798746 |
| 37 | 40.785507 | 87 | 95.901057 |
| 38 | 41.887818 | 88 | 97.003368 |
| 39 | 42.990129 | 89 | 98.105679 |
| 40 | 44.092440 | 90 | 99.207990 |
| 41 | 45.194751 | 91 | 100.310301 |
| 42 | 46.297062 | 92 | 101.412612 |
| 43 | 47.399373 | 93 | 102.514923 |
| 44 | 48.501684 | 94 | 103.617234 |
| 45 | 49.603995 | 95 | 104.719545 |
| 46 | 50.706306 | 96 | 105.821856 |
| 47 | 51.808617 | 97 | 106.924167 |
| 48 | 52.910928 | 98 | 108.026478 |
| 49 | 54.013239 | 99 | 109.128789 |
| 50 | 55.115550 | 100 | 110.231100 |

## WEIGHT
### Long Ton to Metric Ton

| lg ton | mt ton | lg ton | mt ton |
|---|---|---|---|
| 1 | 1.016047 | 51 | 51.818397 |
| 2 | 2.032094 | 52 | 52.834444 |
| 3 | 3.048141 | 53 | 53.850491 |
| 4 | 4.064188 | 54 | 54.866538 |
| 5 | 5.080235 | 55 | 55.882585 |
| 6 | 6.096282 | 56 | 56.898632 |
| 7 | 7.112329 | 57 | 57.914679 |
| 8 | 8.128376 | 58 | 58.930726 |
| 9 | 9.144423 | 59 | 59.946773 |
| 10 | 10.160470 | 60 | 60.962820 |
| 11 | 11.176517 | 61 | 61.978867 |
| 12 | 12.192564 | 62 | 62.994914 |
| 13 | 13.208611 | 63 | 64.010961 |
| 14 | 14.224658 | 64 | 65.027008 |
| 15 | 15.240705 | 65 | 66.043055 |
| 16 | 16.256752 | 66 | 67.059102 |
| 17 | 17.272799 | 67 | 68.075149 |
| 18 | 18.288846 | 68 | 69.091196 |
| 19 | 19.304893 | 69 | 70.107243 |
| 20 | 20.320940 | 70 | 71.123290 |
| 21 | 21.336987 | 71 | 72.139337 |
| 22 | 22.353034 | 72 | 73.155384 |
| 23 | 23.369081 | 73 | 74.171431 |
| 24 | 24.385128 | 74 | 75.187478 |
| 25 | 25.401175 | 75 | 76.203525 |
| 26 | 26.417222 | 76 | 77.219572 |
| 27 | 27.433269 | 77 | 78.235619 |
| 28 | 28.449316 | 78 | 79.251666 |
| 29 | 29.465363 | 79 | 80.267713 |
| 30 | 30.481410 | 80 | 81.283760 |
| 31 | 31.497457 | 81 | 82.299807 |
| 32 | 32.513504 | 82 | 83.315854 |
| 33 | 33.529551 | 83 | 84.331901 |
| 34 | 34.545598 | 84 | 85.347948 |
| 35 | 35.561645 | 85 | 86.363995 |
| 36 | 36.577692 | 86 | 87.380042 |
| 37 | 37.593739 | 87 | 88.396089 |
| 38 | 38.609786 | 88 | 89.412136 |
| 39 | 39.625833 | 89 | 90.428183 |
| 40 | 40.641880 | 90 | 91.444230 |
| 41 | 41.657927 | 91 | 92.460277 |
| 42 | 42.673974 | 92 | 93.476324 |
| 43 | 43.690021 | 93 | 94.492371 |
| 44 | 44.706068 | 94 | 95.508418 |
| 45 | 45.722115 | 95 | 96.524465 |
| 46 | 46.738162 | 96 | 97.540512 |
| 47 | 47.754209 | 97 | 98.556559 |
| 48 | 48.770256 | 98 | 99.572606 |
| 49 | 49.786303 | 99 | 100.588653 |
| 50 | 50.802350 | 100 | 101.604700 |

## WEIGHT
### Metric Ton to Long Ton

| mt ton | lg ton | mt ton | lg ton |
|---|---|---|---|
| 1 | .984206 | 51 | 50.194506 |
| 2 | 1.968412 | 52 | 51.178712 |
| 3 | 2.952618 | 53 | 52.162918 |
| 4 | 3.936824 | 54 | 53.147124 |
| 5 | 4.921030 | 55 | 54.131330 |
| 6 | 5.905236 | 56 | 55.115536 |
| 7 | 6.889442 | 57 | 56.099742 |
| 8 | 7.873648 | 58 | 57.083948 |
| 9 | 8.857854 | 59 | 58.068154 |
| 10 | 9.842060 | 60 | 59.052360 |
| 11 | 10.826266 | 61 | 60.036566 |
| 12 | 11.810472 | 62 | 61.020772 |
| 13 | 12.794678 | 63 | 62.004978 |
| 14 | 13.778884 | 64 | 62.989184 |
| 15 | 14.763090 | 65 | 63.973390 |
| 16 | 15.747296 | 66 | 64.957596 |
| 17 | 16.731502 | 67 | 65.941802 |
| 18 | 17.715708 | 68 | 66.926008 |
| 19 | 18.699914 | 69 | 67.910214 |
| 20 | 19.684120 | 70 | 68.894420 |
| 21 | 20.668326 | 71 | 69.878626 |
| 22 | 21.652532 | 72 | 70.862832 |
| 23 | 22.636738 | 73 | 71.847038 |
| 24 | 23.620944 | 74 | 72.831244 |
| 25 | 24.605150 | 75 | 73.815450 |
| 26 | 25.589356 | 76 | 74.799656 |
| 27 | 26.573562 | 77 | 75.783862 |
| 28 | 27.557768 | 78 | 76.768068 |
| 29 | 28.541974 | 79 | 77.752274 |
| 30 | 29.526180 | 80 | 78.736480 |
| 31 | 30.510386 | 81 | 79.720686 |
| 32 | 31.494592 | 82 | 80.704892 |
| 33 | 32.478798 | 83 | 81.689098 |
| 34 | 33.463004 | 84 | 82.673304 |
| 35 | 34.447210 | 85 | 83.657510 |
| 36 | 35.431416 | 86 | 84.641716 |
| 37 | 36.415622 | 87 | 85.625922 |
| 38 | 37.399828 | 88 | 86.610128 |
| 39 | 38.384034 | 89 | 87.594334 |
| 40 | 39.368240 | 90 | 88.578540 |
| 41 | 40.352446 | 91 | 89.562746 |
| 42 | 41.336652 | 92 | 90.546952 |
| 43 | 42.320858 | 93 | 91.531158 |
| 44 | 43.305064 | 94 | 92.515364 |
| 45 | 44.289270 | 95 | 93.499570 |
| 46 | 45.273476 | 96 | 94.483776 |
| 47 | 46.257682 | 97 | 95.467982 |
| 48 | 47.241888 | 98 | 96.452188 |
| 49 | 48.226094 | 99 | 97.436394 |
| 50 | 49.210300 | 100 | 98.420600 |

## DENSITY
## Pounds per Cubic Inch to Grams per Cubic Centimeter

| lb/in³ | g/cm³ | lb/in³ | g/cm³ |
|---|---|---|---|
| 1 | 27.679910 | 51 | 1411.675410 |
| 2 | 55.359820 | 52 | 1439.355320 |
| 3 | 83.039730 | 53 | 1467.035230 |
| 4 | 110.719640 | 54 | 1494.715140 |
| 5 | 138.399550 | 55 | 1522.395050 |
| 6 | 166.079460 | 56 | 1550.074960 |
| 7 | 193.759370 | 57 | 1577.754870 |
| 8 | 221.439280 | 58 | 1605.434780 |
| 9 | 249.119190 | 59 | 1633.114690 |
| 10 | 276.799100 | 60 | 1660.794600 |
| 11 | 304.479010 | 61 | 1688.474510 |
| 12 | 332.158920 | 62 | 1716.154420 |
| 13 | 359.838830 | 63 | 1743.834330 |
| 14 | 387.518740 | 64 | 1771.514240 |
| 15 | 415.198650 | 65 | 1799.194150 |
| 16 | 442.878560 | 66 | 1826.874060 |
| 17 | 470.558470 | 67 | 1854.553970 |
| 18 | 498.238380 | 68 | 1882.233880 |
| 19 | 525.918290 | 69 | 1909.913790 |
| 20 | 553.598200 | 70 | 1937.593700 |
| 21 | 581.278110 | 71 | 1965.273610 |
| 22 | 608.958020 | 72 | 1992.953520 |
| 23 | 636.637930 | 73 | 2020.633430 |
| 24 | 664.317840 | 74 | 2048.313340 |
| 25 | 691.997750 | 75 | 2075.993250 |
| 26 | 719.677660 | 76 | 2103.673160 |
| 27 | 747.357570 | 77 | 2131.353070 |
| 28 | 775.037480 | 78 | 2159.032980 |
| 29 | 802.717390 | 79 | 2186.712890 |
| 30 | 830.397300 | 80 | 2214.392800 |
| 31 | 858.077210 | 81 | 2242.072710 |
| 32 | 885.757120 | 82 | 2269.752620 |
| 33 | 913.437030 | 83 | 2297.432530 |
| 34 | 941.116940 | 84 | 2325.112440 |
| 35 | 968.796850 | 85 | 2352.792350 |
| 36 | 996.476760 | 86 | 2380.472260 |
| 37 | 1024.156670 | 87 | 2408.152170 |
| 38 | 1051.836580 | 88 | 2435.832080 |
| 39 | 1079.516490 | 89 | 2463.511990 |
| 40 | 1107.196400 | 90 | 2491.191900 |
| 41 | 1134.876310 | 91 | 2518.871810 |
| 42 | 1162.556220 | 92 | 2546.551720 |
| 43 | 1190.236130 | 93 | 2574.231630 |
| 44 | 1217.916040 | 94 | 2601.911540 |
| 45 | 1245.595950 | 95 | 2629.591450 |
| 46 | 1273.275860 | 96 | 2657.271360 |
| 47 | 1300.955770 | 97 | 2684.951270 |
| 48 | 1328.635680 | 98 | 2712.631180 |
| 49 | 1356.315590 | 99 | 2740.311090 |
| 50 | 1383.995500 | 100 | 2767.991000 |

## DENSITY
## Grams per Cubic Centimeter to Pounds per Cubic Inch

| g/cm³ | lb/in³ | g/cm³ | lb/in³ |
|---|---|---|---|
| 1 | .036127 | 51 | 1.842477 |
| 2 | .072254 | 52 | 1.878604 |
| 3 | .108381 | 53 | 1.914731 |
| 4 | .144508 | 54 | 1.950858 |
| 5 | .180635 | 55 | 1.986985 |
| 6 | .216762 | 56 | 2.023112 |
| 7 | .252889 | 57 | 2.059239 |
| 8 | .289016 | 58 | 2.095366 |
| 9 | .325143 | 59 | 2.131493 |
| 10 | .361270 | 60 | 2.167620 |
| 11 | .397397 | 61 | 2.203747 |
| 12 | .433524 | 62 | 2.239874 |
| 13 | .469651 | 63 | 2.276001 |
| 14 | .505778 | 64 | 2.312128 |
| 15 | .541905 | 65 | 2.348255 |
| 16 | .578032 | 66 | 2.384382 |
| 17 | .614159 | 67 | 2.420509 |
| 18 | .650286 | 68 | 2.456636 |
| 19 | .686413 | 69 | 2.492763 |
| 20 | .722540 | 70 | 2.528890 |
| 21 | .758667 | 71 | 2.565017 |
| 22 | .794794 | 72 | 2.601144 |
| 23 | .830921 | 73 | 2.637271 |
| 24 | .867048 | 74 | 2.673398 |
| 25 | .903175 | 75 | 2.709525 |
| 26 | .939302 | 76 | 2.745652 |
| 27 | .975429 | 77 | 2.781779 |
| 28 | 1.011556 | 78 | 2.817906 |
| 29 | 1.047683 | 79 | 2.854033 |
| 30 | 1.083810 | 80 | 2.890160 |
| 31 | 1.119937 | 81 | 2.926287 |
| 32 | 1.156064 | 82 | 2.962414 |
| 33 | 1.192191 | 83 | 2.998541 |
| 34 | 1.228318 | 84 | 3.034668 |
| 35 | 1.264445 | 85 | 3.070795 |
| 36 | 1.300572 | 86 | 3.106922 |
| 37 | 1.336699 | 87 | 3.143049 |
| 38 | 1.372826 | 88 | 3.179176 |
| 39 | 1.408953 | 89 | 3.215303 |
| 40 | 1.445080 | 90 | 3.251430 |
| 41 | 1.481207 | 91 | 3.287557 |
| 42 | 1.517334 | 92 | 3.323684 |
| 43 | 1.553461 | 93 | 3.359811 |
| 44 | 1.589588 | 94 | 3.395938 |
| 45 | 1.625715 | 95 | 3.432065 |
| 46 | 1.661842 | 96 | 3.468192 |
| 47 | 1.697969 | 97 | 3.504319 |
| 48 | 1.734096 | 98 | 3.540446 |
| 49 | 1.770223 | 99 | 3.576573 |
| 50 | 1.806350 | 100 | 3.612700 |

## DENSITY
## Pounds per Cubic Inch
## to
## Kilograms per Cubic Centimeter

| lb/in³ | kg/cm³ | lb/in³ | kg/cm³ |
|---|---|---|---|
| 1 | .027680 | 51 | 1.411680 |
| 2 | .055360 | 52 | 1.439360 |
| 3 | .083040 | 53 | 1.467040 |
| 4 | .110720 | 54 | 1.494720 |
| 5 | .138400 | 55 | 1.522400 |
| 6 | .166080 | 56 | 1.550080 |
| 7 | .193760 | 57 | 1.577760 |
| 8 | .221440 | 58 | 1.605440 |
| 9 | .249120 | 59 | 1.633120 |
| 10 | .276800 | 60 | 1.660800 |
| 11 | .304480 | 61 | 1.688480 |
| 12 | .332160 | 62 | 1.716160 |
| 13 | .359840 | 63 | 1.743840 |
| 14 | .387520 | 64 | 1.771520 |
| 15 | .415200 | 65 | 1.799200 |
| 16 | .442880 | 66 | 1.826880 |
| 17 | .470560 | 67 | 1.854560 |
| 18 | .498240 | 68 | 1.882240 |
| 19 | .525920 | 69 | 1.909920 |
| 20 | .553600 | 70 | 1.937600 |
| 21 | .581280 | 71 | 1.965280 |
| 22 | .608960 | 72 | 1.992960 |
| 23 | .636640 | 73 | 2.020640 |
| 24 | .664320 | 74 | 2.048320 |
| 25 | .692000 | 75 | 2.076000 |
| 26 | .719680 | 76 | 2.103680 |
| 27 | .747360 | 77 | 2.131360 |
| 28 | .775040 | 78 | 2.159040 |
| 29 | .802720 | 79 | 2.186720 |
| 30 | .830400 | 80 | 2.214400 |
| 31 | .858080 | 81 | 2.242080 |
| 32 | .885760 | 82 | 2.269760 |
| 33 | .913440 | 83 | 2.297440 |
| 34 | .941120 | 84 | 2.325120 |
| 35 | .968800 | 85 | 2.352800 |
| 36 | .996480 | 86 | 2.380480 |
| 37 | 1.024160 | 87 | 2.408160 |
| 38 | 1.051840 | 88 | 2.435840 |
| 39 | 1.079520 | 89 | 2.463520 |
| 40 | 1.107200 | 90 | 2.491200 |
| 41 | 1.134880 | 91 | 2.518880 |
| 42 | 1.162560 | 92 | 2.546560 |
| 43 | 1.190240 | 93 | 2.574240 |
| 44 | 1.217920 | 94 | 2.601920 |
| 45 | 1.245600 | 95 | 2.629600 |
| 46 | 1.273280 | 96 | 2.657280 |
| 47 | 1.300960 | 97 | 2.684960 |
| 48 | 1.328640 | 98 | 2.712640 |
| 49 | 1.356320 | 99 | 2.740320 |
| 50 | 1.384000 | 100 | 2.768000 |

# DENSITY
## Kilograms per Cubic Centimeter to Pounds per Cubic Inch

| kg/cm³ | lb/in³ | kg/cm³ | lb/in³ |
|---|---|---|---|
| 1 | 36.127200 | 51 | 1842.487200 |
| 2 | 72.254400 | 52 | 1878.614400 |
| 3 | 108.381600 | 53 | 1914.741600 |
| 4 | 144.508800 | 54 | 1950.868800 |
| 5 | 180.636000 | 55 | 1986.996000 |
| 6 | 216.763200 | 56 | 2023.123200 |
| 7 | 252.890400 | 57 | 2059.250400 |
| 8 | 289.017600 | 58 | 2095.377600 |
| 9 | 325.144800 | 59 | 2131.504800 |
| 10 | 361.272000 | 60 | 2167.632000 |
| 11 | 397.399200 | 61 | 2203.759200 |
| 12 | 433.526400 | 62 | 2239.886400 |
| 13 | 469.653600 | 63 | 2276.013600 |
| 14 | 505.780800 | 64 | 2312.140800 |
| 15 | 541.908000 | 65 | 2348.268000 |
| 16 | 578.035200 | 66 | 2384.395200 |
| 17 | 614.162400 | 67 | 2420.522400 |
| 18 | 650.289600 | 68 | 2456.649600 |
| 19 | 686.416800 | 69 | 2492.776800 |
| 20 | 722.544000 | 70 | 2528.904000 |
| 21 | 758.671200 | 71 | 2565.031200 |
| 22 | 794.798400 | 72 | 2601.158400 |
| 23 | 830.925600 | 73 | 2637.285600 |
| 24 | 867.052800 | 74 | 2673.412800 |
| 25 | 903.180000 | 75 | 2709.540000 |
| 26 | 939.307200 | 76 | 2745.667200 |
| 27 | 975.434400 | 77 | 2781.794400 |
| 28 | 1011.561600 | 78 | 2817.921600 |
| 29 | 1047.688800 | 79 | 2854.048800 |
| 30 | 1083.816000 | 80 | 2890.176000 |
| 31 | 1119.943200 | 81 | 2926.303200 |
| 32 | 1156.070400 | 82 | 2962.430400 |
| 33 | 1192.197600 | 83 | 2998.557600 |
| 34 | 1228.324800 | 84 | 3034.684800 |
| 35 | 1264.452000 | 85 | 3070.812000 |
| 36 | 1300.579200 | 86 | 3106.939200 |
| 37 | 1336.706400 | 87 | 3143.066400 |
| 38 | 1372.833600 | 88 | 3179.193600 |
| 39 | 1408.960800 | 89 | 3215.320800 |
| 40 | 1445.088000 | 90 | 3251.448000 |
| 41 | 1481.215200 | 91 | 3287.575200 |
| 42 | 1517.342400 | 92 | 3323.702400 |
| 43 | 1553.469600 | 93 | 3359.829600 |
| 44 | 1589.596800 | 94 | 3395.956800 |
| 45 | 1625.724000 | 95 | 3432.084000 |
| 46 | 1661.851200 | 96 | 3468.211200 |
| 47 | 1697.978400 | 97 | 3504.338400 |
| 48 | 1734.105600 | 98 | 3540.465600 |
| 49 | 1770.232800 | 99 | 3576.592800 |
| 50 | 1806.360000 | 100 | 3612.720000 |

## DENSITY
## Pounds per Cubic Foot
## to
## Kilograms per Cubic Meter

| lb/ft³ | kg/m³ | lb/ft³ | kg/m³ |
|---|---|---|---|
| 1 | 16.018460 | 51 | 816.941460 |
| 2 | 32.036920 | 52 | 832.959920 |
| 3 | 48.055380 | 53 | 848.978380 |
| 4 | 64.073840 | 54 | 864.996840 |
| 5 | 80.092300 | 55 | 881.015300 |
| 6 | 96.110760 | 56 | 897.033760 |
| 7 | 112.129220 | 57 | 913.052220 |
| 8 | 128.147680 | 58 | 929.070680 |
| 9 | 144.166140 | 59 | 945.089140 |
| 10 | 160.184600 | 60 | 961.107600 |
| 11 | 176.203060 | 61 | 977.126060 |
| 12 | 192.221520 | 62 | 993.144520 |
| 13 | 208.239980 | 63 | 1009.162980 |
| 14 | 224.258440 | 64 | 1025.181440 |
| 15 | 240.276900 | 65 | 1041.199900 |
| 16 | 256.295360 | 66 | 1057.218360 |
| 17 | 272.313820 | 67 | 1073.236820 |
| 18 | 288.332280 | 68 | 1089.255280 |
| 19 | 304.350740 | 69 | 1105.273740 |
| 20 | 320.369200 | 70 | 1121.292200 |
| 21 | 336.387660 | 71 | 1137.310660 |
| 22 | 352.406120 | 72 | 1153.329120 |
| 23 | 368.424580 | 73 | 1169.347580 |
| 24 | 384.443040 | 74 | 1185.366040 |
| 25 | 400.461500 | 75 | 1201.384500 |
| 26 | 416.479960 | 76 | 1217.402960 |
| 27 | 432.498420 | 77 | 1233.421420 |
| 28 | 448.516880 | 78 | 1249.439880 |
| 29 | 464.535340 | 79 | 1265.458340 |
| 30 | 480.553800 | 80 | 1281.476800 |
| 31 | 496.572260 | 81 | 1297.495260 |
| 32 | 512.590720 | 82 | 1313.513720 |
| 33 | 528.609180 | 83 | 1329.532180 |
| 34 | 544.627640 | 84 | 1345.550640 |
| 35 | 560.646100 | 85 | 1361.569100 |
| 36 | 576.664560 | 86 | 1377.587560 |
| 37 | 592.683020 | 87 | 1393.606020 |
| 38 | 608.701480 | 88 | 1409.624480 |
| 39 | 624.719940 | 89 | 1425.642940 |
| 40 | 640.738400 | 90 | 1441.661400 |
| 41 | 656.756860 | 91 | 1457.679860 |
| 42 | 672.775320 | 92 | 1473.698320 |
| 43 | 688.793780 | 93 | 1489.716780 |
| 44 | 704.812240 | 94 | 1505.735240 |
| 45 | 720.830700 | 95 | 1521.753700 |
| 46 | 736.849160 | 96 | 1537.772160 |
| 47 | 752.867620 | 97 | 1553.790620 |
| 48 | 768.886080 | 98 | 1569.809080 |
| 49 | 784.904540 | 99 | 1585.827540 |
| 50 | 800.923000 | 100 | 1601.846000 |

## DENSITY
### Kilograms per Cubic Meter to Pounds per Cubic Foot

| kg/m³ | lb/ft³ | kg/m³ | lb/ft³ |
|---|---|---|---|
| 1 | .062428 | 51 | 3.183828 |
| 2 | .124856 | 52 | 3.246256 |
| 3 | .187284 | 53 | 3.308684 |
| 4 | .249712 | 54 | 3.371112 |
| 5 | .312140 | 55 | 3.433540 |
| 6 | .374568 | 56 | 3.495968 |
| 7 | .436996 | 57 | 3.558396 |
| 8 | .499424 | 58 | 3.620824 |
| 9 | .561852 | 59 | 3.683252 |
| 10 | .624280 | 60 | 3.745680 |
| 11 | .686708 | 61 | 3.808108 |
| 12 | .749136 | 62 | 3.870536 |
| 13 | .811564 | 63 | 3.932964 |
| 14 | .873992 | 64 | 3.995392 |
| 15 | .936420 | 65 | 4.057820 |
| 16 | .998848 | 66 | 4.120248 |
| 17 | 1.061276 | 67 | 4.182676 |
| 18 | 1.123704 | 68 | 4.245104 |
| 19 | 1.186132 | 69 | 4.307532 |
| 20 | 1.248560 | 70 | 4.369960 |
| 21 | 1.310988 | 71 | 4.432388 |
| 22 | 1.373416 | 72 | 4.494816 |
| 23 | 1.435844 | 73 | 4.557244 |
| 24 | 1.498272 | 74 | 4.619672 |
| 25 | 1.560700 | 75 | 4.682100 |
| 26 | 1.623128 | 76 | 4.744528 |
| 27 | 1.685556 | 77 | 4.806956 |
| 28 | 1.747984 | 78 | 4.869384 |
| 29 | 1.810412 | 79 | 4.931812 |
| 30 | 1.872840 | 80 | 4.994240 |
| 31 | 1.935268 | 81 | 5.056668 |
| 32 | 1.997696 | 82 | 5.119096 |
| 33 | 2.060124 | 83 | 5.181524 |
| 34 | 2.122552 | 84 | 5.243952 |
| 35 | 2.184980 | 85 | 5.306380 |
| 36 | 2.247408 | 86 | 5.368808 |
| 37 | 2.309836 | 87 | 5.431236 |
| 38 | 2.372264 | 88 | 5.493664 |
| 39 | 2.434692 | 89 | 5.556092 |
| 40 | 2.497120 | 90 | 5.618520 |
| 41 | 2.559548 | 91 | 5.680948 |
| 42 | 2.621976 | 92 | 5.743376 |
| 43 | 2.684404 | 93 | 5.805804 |
| 44 | 2.746832 | 94 | 5.868232 |
| 45 | 2.809260 | 95 | 5.930660 |
| 46 | 2.871688 | 96 | 5.993088 |
| 47 | 2.934116 | 97 | 6.055516 |
| 48 | 2.996544 | 98 | 6.117944 |
| 49 | 3.058972 | 99 | 6.180372 |
| 50 | 3.121400 | 100 | 6.242800 |

## DENSITY
### Pounds per Cubic Foot to Grams per Cubic Centimeter

| lb/ft³ | g/cm³ | lb/ft³ | g/cm³ |
|---|---|---|---|
| 1 | .016018 | 51 | .816918 |
| 2 | .032036 | 52 | .832936 |
| 3 | .048054 | 53 | .848954 |
| 4 | .064072 | 54 | .864972 |
| 5 | .080090 | 55 | .880990 |
| 6 | .096108 | 56 | .897008 |
| 7 | .112126 | 57 | .913026 |
| 8 | .128144 | 58 | .929044 |
| 9 | .144162 | 59 | .945062 |
| 10 | .160180 | 60 | .961080 |
| 11 | .176198 | 61 | .977098 |
| 12 | .192216 | 62 | .993116 |
| 13 | .208234 | 63 | 1.009134 |
| 14 | .224252 | 64 | 1.025152 |
| 15 | .240270 | 65 | 1.041170 |
| 16 | .256288 | 66 | 1.057188 |
| 17 | .272306 | 67 | 1.073206 |
| 18 | .288324 | 68 | 1.089224 |
| 19 | .304342 | 69 | 1.105242 |
| 20 | .320360 | 70 | 1.121260 |
| 21 | .336378 | 71 | 1.137278 |
| 22 | .352396 | 72 | 1.153296 |
| 23 | .368414 | 73 | 1.169314 |
| 24 | .384432 | 74 | 1.185332 |
| 25 | .400450 | 75 | 1.201350 |
| 26 | .416468 | 76 | 1.217368 |
| 27 | .432486 | 77 | 1.233386 |
| 28 | .448504 | 78 | 1.249404 |
| 29 | .464522 | 79 | 1.265422 |
| 30 | .480540 | 80 | 1.281440 |
| 31 | .496558 | 81 | 1.297458 |
| 32 | .512576 | 82 | 1.313476 |
| 33 | .528594 | 83 | 1.329494 |
| 34 | .544612 | 84 | 1.345512 |
| 35 | .560630 | 85 | 1.361530 |
| 36 | .576648 | 86 | 1.377548 |
| 37 | .592666 | 87 | 1.393566 |
| 38 | .608684 | 88 | 1.409584 |
| 39 | .624702 | 89 | 1.425602 |
| 40 | .640720 | 90 | 1.441620 |
| 41 | .656738 | 91 | 1.457638 |
| 42 | .672756 | 92 | 1.473656 |
| 43 | .688774 | 93 | 1.489674 |
| 44 | .704792 | 94 | 1.505692 |
| 45 | .720810 | 95 | 1.521710 |
| 46 | .736828 | 96 | 1.537728 |
| 47 | .752846 | 97 | 1.553746 |
| 48 | .768864 | 98 | 1.569764 |
| 49 | .784882 | 99 | 1.585782 |
| 50 | .800900 | 100 | 1.601800 |

## DENSITY
### Grams per Cubic Centimeter to Pounds per Cubic Foot

| g/cm³ | lb/ft³ | g/cm³ | lb/ft³ |
|---|---|---|---|
| 1 | 62.427950 | 51 | 3183.825450 |
| 2 | 124.855900 | 52 | 3246.253400 |
| 3 | 187.283850 | 53 | 3308.681350 |
| 4 | 249.711800 | 54 | 3371.109300 |
| 5 | 312.139750 | 55 | 3433.537250 |
| 6 | 374.567700 | 56 | 3495.965200 |
| 7 | 436.995650 | 57 | 3558.393150 |
| 8 | 499.423600 | 58 | 3620.821100 |
| 9 | 561.851550 | 59 | 3683.249050 |
| 10 | 624.279500 | 60 | 3745.677000 |
| 11 | 686.707450 | 61 | 3808.104950 |
| 12 | 749.135400 | 62 | 3870.532900 |
| 13 | 811.563350 | 63 | 3932.960850 |
| 14 | 873.991300 | 64 | 3995.388800 |
| 15 | 936.419250 | 65 | 4057.816750 |
| 16 | 998.847200 | 66 | 4120.244700 |
| 17 | 1061.275150 | 67 | 4182.672650 |
| 18 | 1123.703100 | 68 | 4245.100600 |
| 19 | 1186.131050 | 69 | 4307.528550 |
| 20 | 1248.559000 | 70 | 4369.956500 |
| 21 | 1310.986950 | 71 | 4432.384450 |
| 22 | 1373.414900 | 72 | 4494.812400 |
| 23 | 1435.842850 | 73 | 4557.240350 |
| 24 | 1498.270800 | 74 | 4619.668300 |
| 25 | 1560.698750 | 75 | 4682.096250 |
| 26 | 1623.126700 | 76 | 4744.524200 |
| 27 | 1685.554650 | 77 | 4806.952150 |
| 28 | 1747.982600 | 78 | 4869.380100 |
| 29 | 1810.410550 | 79 | 4931.808050 |
| 30 | 1872.838500 | 80 | 4994.236000 |
| 31 | 1935.266450 | 81 | 5056.663950 |
| 32 | 1997.694400 | 82 | 5119.091900 |
| 33 | 2060.122350 | 83 | 5181.519850 |
| 34 | 2122.550300 | 84 | 5243.947800 |
| 35 | 2184.978250 | 85 | 5306.375750 |
| 36 | 2247.406200 | 86 | 5368.803700 |
| 37 | 2309.834150 | 87 | 5431.231650 |
| 38 | 2372.262100 | 88 | 5493.659600 |
| 39 | 2434.690050 | 89 | 5556.087550 |
| 40 | 2497.118000 | 90 | 5618.515500 |
| 41 | 2559.545950 | 91 | 5680.943450 |
| 42 | 2621.973900 | 92 | 5743.371400 |
| 43 | 2684.401850 | 93 | 5805.799350 |
| 44 | 2746.829800 | 94 | 5868.227300 |
| 45 | 2809.257750 | 95 | 5930.655250 |
| 46 | 2871.685700 | 96 | 5993.083200 |
| 47 | 2934.113650 | 97 | 6055.511150 |
| 48 | 2996.541600 | 98 | 6117.939100 |
| 49 | 3058.969550 | 99 | 6180.367050 |
| 50 | 3121.397500 | 100 | 6242.795000 |

## TEMPERATURE
### Fahrenheit to Celsius

| °F | °C | °F | °C |
|---|---|---|---|
| -100 | -73.333260 | -50 | -45.555510 |
| -99 | -72.777705 | -49 | -44.999955 |
| -98 | -72.222150 | -48 | -44.444400 |
| -97 | -71.666595 | -47 | -43.888845 |
| -96 | -71.111040 | -46 | -43.333290 |
| -95 | -70.555485 | -45 | -42.777735 |
| -94 | -69.999930 | -44 | -42.222180 |
| -93 | -69.444375 | -43 | -41.666625 |
| -92 | -68.888820 | -42 | -41.111070 |
| -91 | -68.333265 | -41 | -40.555515 |
| -90 | -67.777710 | -40 | -39.999960 |
| -89 | -67.222155 | -39 | -39.444405 |
| -88 | -66.666600 | -38 | -38.888850 |
| -87 | -66.111045 | -37 | -38.333295 |
| -86 | -65.555490 | -36 | -37.777740 |
| -85 | -64.999935 | -35 | -37.222185 |
| -84 | -64.444380 | -34 | -36.666630 |
| -83 | -63.888825 | -33 | -36.111075 |
| -82 | -63.333270 | -32 | -35.555520 |
| -81 | -62.777715 | -31 | -34.999965 |
| -80 | -62.222160 | -30 | -34.444410 |
| -79 | -61.666605 | -29 | -33.888855 |
| -78 | -61.111050 | -28 | -33.333300 |
| -77 | -60.555495 | -27 | -32.777745 |
| -76 | -59.999940 | -26 | -32.222190 |
| -75 | -59.444385 | -25 | -31.666635 |
| -74 | -58.888830 | -24 | -31.111080 |
| -73 | -58.333275 | -23 | -30.555525 |
| -72 | -57.777720 | -22 | -29.999970 |
| -71 | -57.222165 | -21 | -29.444415 |
| -70 | -56.666610 | -20 | -28.888860 |
| -69 | -56.111055 | -19 | -28.333305 |
| -68 | -55.555500 | -18 | -27.777750 |
| -67 | -54.999945 | -17 | -27.222195 |
| -66 | -54.444390 | -16 | -26.666640 |
| -65 | -53.888835 | -15 | -26.111085 |
| -64 | -53.333280 | -14 | -25.555530 |
| -63 | -52.777725 | -13 | -24.999975 |
| -62 | -52.222170 | -12 | -24.444420 |
| -61 | -51.666615 | -11 | -23.888865 |
| -60 | -51.111060 | -10 | -23.333310 |
| -59 | -50.555505 | -9 | -22.777755 |
| -58 | -49.999950 | -8 | -22.222200 |
| -57 | -49.444395 | -7 | -21.666645 |
| -56 | -48.888840 | -6 | -21.111090 |
| -55 | -48.333285 | -5 | -20.555535 |
| -54 | -47.777730 | -4 | -19.999980 |
| -53 | -47.222175 | -3 | -19.444425 |
| -52 | -46.666620 | -2 | -18.888870 |
| -51 | -46.111065 | -1 | -18.333315 |
| | | 0 | -17.777760 |

## TEMPERATURE
### Fahrenheit to Celsius (cont'd)

| °F | °C | °F | °C |
|---|---|---|---|
| 1 | -17.222205 | 51 | 10.555545 |
| 2 | -16.666650 | 52 | 11.111100 |
| 3 | -16.111095 | 53 | 11.666655 |
| 4 | -15.555540 | 54 | 12.222210 |
| 5 | -14.999985 | 55 | 12.777765 |
| 6 | -14.444430 | 56 | 13.333320 |
| 7 | -13.888875 | 57 | 13.888875 |
| 8 | -13.333320 | 58 | 14.444430 |
| 9 | -12.777765 | 59 | 14.999985 |
| 10 | -12.222210 | 60 | 15.555540 |
| 11 | -11.666655 | 61 | 16.111095 |
| 12 | -11.111100 | 62 | 16.666650 |
| 13 | -10.555545 | 63 | 17.222205 |
| 14 | -9.999990 | 64 | 17.777760 |
| 15 | -9.444435 | 65 | 18.333315 |
| 16 | -8.888880 | 66 | 18.888870 |
| 17 | -8.333325 | 67 | 19.444425 |
| 18 | -7.777770 | 68 | 19.999980 |
| 19 | -7.222215 | 69 | 20.555535 |
| 20 | -6.666660 | 70 | 21.111090 |
| 21 | -6.111105 | 71 | 21.666645 |
| 22 | -5.555550 | 72 | 22.222200 |
| 23 | -4.999995 | 73 | 22.777755 |
| 24 | -4.444440 | 74 | 23.333310 |
| 25 | -3.888885 | 75 | 23.888865 |
| 26 | -3.333330 | 76 | 24.444420 |
| 27 | -2.777775 | 77 | 24.999975 |
| 28 | -2.222220 | 78 | 25.555530 |
| 29 | -1.666665 | 79 | 26.111085 |
| 30 | -1.111110 | 80 | 26.666640 |
| 31 | -.555555 | 81 | 27.222195 |
| 32 | .000000 | 82 | 27.777750 |
| 33 | .555555 | 83 | 28.333305 |
| 34 | 1.111110 | 84 | 28.888860 |
| 35 | 1.666665 | 85 | 29.444415 |
| 36 | 2.222220 | 86 | 29.999970 |
| 37 | 2.777775 | 87 | 30.555525 |
| 38 | 3.333330 | 88 | 31.111080 |
| 39 | 3.888885 | 89 | 31.666635 |
| 40 | 4.444440 | 90 | 32.222190 |
| 41 | 4.999995 | 91 | 32.777745 |
| 42 | 5.555550 | 92 | 33.333300 |
| 43 | 6.111105 | 93 | 33.888855 |
| 44 | 6.666660 | 94 | 34.444410 |
| 45 | 7.222215 | 95 | 34.999965 |
| 46 | 7.777770 | 96 | 35.555520 |
| 47 | 8.333325 | 97 | 36.111075 |
| 48 | 8.888880 | 98 | 36.666630 |
| 49 | 9.444435 | 99 | 37.222185 |
| 50 | 9.999990 | 100 | 37.777740 |

## TEMPERATURE
### Fahrenheit to Celsius (cont'd)

| °F | °C | °F | °C |
|---|---|---|---|
| 105 | 40.555515 | 305 | 151.666515 |
| 110 | 43.333290 | 310 | 154.444290 |
| 115 | 46.111065 | 315 | 157.222065 |
| 120 | 48.888840 | 320 | 159.999840 |
| 125 | 51.666615 | 325 | 162.777615 |
| 130 | 54.444390 | 330 | 165.555390 |
| 135 | 57.222165 | 335 | 168.333165 |
| 140 | 59.999940 | 340 | 171.110940 |
| 145 | 62.777715 | 345 | 173.888715 |
| 150 | 65.555490 | 350 | 176.666490 |
| 155 | 68.333265 | 355 | 179.444265 |
| 160 | 71.111040 | 360 | 182.222040 |
| 165 | 73.888815 | 365 | 184.999815 |
| 170 | 76.666590 | 370 | 187.777590 |
| 175 | 79.444365 | 375 | 190.555365 |
| 180 | 82.222140 | 380 | 193.333140 |
| 185 | 84.999915 | 385 | 196.110915 |
| 190 | 87.777690 | 390 | 198.888690 |
| 195 | 90.555465 | 395 | 201.666465 |
| 200 | 93.333240 | 400 | 204.444240 |
| 205 | 96.111015 | 405 | 207.222015 |
| 210 | 98.888790 | 410 | 209.999790 |
| 215 | 101.666565 | 415 | 212.777565 |
| 220 | 104.444340 | 420 | 215.555340 |
| 225 | 107.222115 | 425 | 218.333115 |
| 230 | 109.999890 | 430 | 221.110890 |
| 235 | 112.777665 | 435 | 223.888665 |
| 240 | 115.555440 | 440 | 226.666440 |
| 245 | 118.333215 | 445 | 229.444215 |
| 250 | 121.110990 | 450 | 232.221990 |
| 255 | 123.888765 | 455 | 234.999765 |
| 260 | 126.666540 | 460 | 237.777540 |
| 265 | 129.444315 | 465 | 240.555315 |
| 270 | 132.222090 | 470 | 243.333090 |
| 275 | 134.999865 | 475 | 246.110865 |
| 280 | 137.777640 | 480 | 248.888640 |
| 285 | 140.555415 | 485 | 251.666415 |
| 290 | 143.333190 | 490 | 254.444190 |
| 295 | 146.110965 | 495 | 257.221965 |
| 300 | 148.888740 | 500 | 259.999740 |

## TEMPERATURE
### Fahrenheit to Celsius (cont'd)

| °F | °C | °F | °C |
|---|---|---|---|
| 510 | 265.555290 | 910 | 487.777290 |
| 520 | 271.110840 | 920 | 493.332840 |
| 530 | 276.666390 | 930 | 498.888390 |
| 540 | 282.221940 | 940 | 504.443940 |
| 550 | 287.777490 | 950 | 509.999490 |
| 560 | 293.333040 | 960 | 515.555040 |
| 570 | 298.888590 | 970 | 521.110590 |
| 580 | 304.444140 | 980 | 526.666140 |
| 590 | 309.999690 | 990 | 532.221690 |
| 600 | 315.555240 | 1000 | 537.777240 |
| 610 | 321.110790 | 1010 | 543.332790 |
| 620 | 326.666340 | 1020 | 548.888340 |
| 630 | 332.221890 | 1030 | 554.443890 |
| 640 | 337.777440 | 1040 | 559.999440 |
| 650 | 343.332990 | 1050 | 565.554990 |
| 660 | 348.888540 | 1060 | 571.110540 |
| 670 | 354.444090 | 1070 | 576.666090 |
| 680 | 359.999640 | 1080 | 582.221640 |
| 690 | 365.555190 | 1090 | 587.777190 |
| 700 | 371.110740 | 1100 | 593.332740 |
| 710 | 376.666290 | 1110 | 598.888290 |
| 720 | 382.221840 | 1120 | 604.443840 |
| 730 | 387.777390 | 1130 | 609.999390 |
| 740 | 393.332940 | 1140 | 615.554940 |
| 750 | 398.888490 | 1150 | 621.110490 |
| 760 | 404.444040 | 1160 | 626.666040 |
| 770 | 409.999590 | 1170 | 632.221590 |
| 780 | 415.555140 | 1180 | 637.777140 |
| 790 | 421.110690 | 1190 | 643.332690 |
| 800 | 426.666240 | 1200 | 648.888240 |
| 810 | 432.221790 | 1210 | 654.443790 |
| 820 | 437.777340 | 1220 | 659.999340 |
| 830 | 443.332890 | 1230 | 665.554890 |
| 840 | 448.888440 | 1240 | 671.110440 |
| 850 | 454.443990 | 1250 | 676.665990 |
| 860 | 459.999540 | 1260 | 682.221540 |
| 870 | 465.555090 | 1270 | 687.777090 |
| 880 | 471.110640 | 1280 | 693.332640 |
| 890 | 476.666190 | 1290 | 698.888190 |
| 900 | 482.221740 | 1300 | 704.443740 |

## TEMPERATURE
### Fahrenheit to Celsius (cont'd)

| °F | °C | °F | °C |
|---|---|---|---|
| 1310 | 709.999290 | 1710 | 932.221290 |
| 1320 | 715.554840 | 1720 | 937.776840 |
| 1330 | 721.110390 | 1730 | 943.332390 |
| 1340 | 726.665940 | 1740 | 948.887940 |
| 1350 | 732.221490 | 1750 | 954.443490 |
| 1360 | 737.777040 | 1760 | 959.999040 |
| 1370 | 743.332590 | 1770 | 965.554590 |
| 1380 | 748.888140 | 1780 | 971.110140 |
| 1390 | 754.443690 | 1790 | 976.665690 |
| 1400 | 759.999240 | 1800 | 982.221240 |
| 1410 | 765.554790 | 1810 | 987.776790 |
| 1420 | 771.110340 | 1820 | 993.332340 |
| 1430 | 776.665890 | 1830 | 998.887890 |
| 1440 | 782.221440 | 1840 | 1004.443440 |
| 1450 | 787.776990 | 1850 | 1009.998990 |
| 1460 | 793.332540 | 1860 | 1015.554540 |
| 1470 | 798.888090 | 1870 | 1021.110090 |
| 1480 | 804.443640 | 1880 | 1026.665640 |
| 1490 | 809.999190 | 1890 | 1032.221190 |
| 1500 | 815.554740 | 1900 | 1037.776740 |
| 1510 | 821.110290 | 1910 | 1043.332290 |
| 1520 | 826.665840 | 1920 | 1048.887840 |
| 1530 | 832.221390 | 1930 | 1054.443390 |
| 1540 | 837.776940 | 1940 | 1059.998940 |
| 1550 | 843.332490 | 1950 | 1065.554490 |
| 1560 | 848.888040 | 1960 | 1071.110040 |
| 1570 | 854.443590 | 1970 | 1076.665590 |
| 1580 | 859.999140 | 1980 | 1082.221140 |
| 1590 | 865.554690 | 1990 | 1087.776690 |
| 1600 | 871.110240 | 2000 | 1093.332240 |
| 1610 | 876.665790 | | |
| 1620 | 882.221340 | | |
| 1630 | 887.776890 | | |
| 1640 | 893.332440 | | |
| 1650 | 898.887990 | | |
| 1660 | 904.443540 | | |
| 1670 | 909.999090 | | |
| 1680 | 915.554640 | | |
| 1690 | 921.110190 | | |
| 1700 | 926.665740 | | |

## TEMPERATURE
### Celsius to Fahrenheit

| °C | °F |
|---|---|
| -100 | -147.960000 |
| -99 | -146.160000 |
| -98 | -144.360000 |
| -97 | -142.560000 |
| -96 | -140.760000 |
| -95 | -138.960000 |
| -94 | -137.160000 |
| -93 | -135.360000 |
| -92 | -133.560000 |
| -91 | -131.760000 |
| -90 | -129.960000 |
| -89 | -128.160000 |
| -88 | -126.360000 |
| -87 | -124.560000 |
| -86 | -122.760000 |
| -85 | -120.960000 |
| -84 | -119.160000 |
| -83 | -117.360000 |
| -82 | -115.560000 |
| -81 | -113.760000 |
| -80 | -111.960000 |
| -79 | -110.160000 |
| -78 | -108.360000 |
| -77 | -106.560000 |
| -76 | -104.760000 |
| -75 | -102.960000 |
| -74 | -101.160000 |
| -73 | -99.360000 |
| -72 | -97.560000 |
| -71 | -95.760000 |
| -70 | -93.960000 |
| -69 | -92.160000 |
| -68 | -90.360000 |
| -67 | -88.560000 |
| -66 | -86.760000 |
| -65 | -84.960000 |
| -64 | -83.160000 |
| -63 | -81.360000 |
| -62 | -79.560000 |
| -61 | -77.760000 |
| -60 | -75.960000 |
| -59 | -74.160000 |
| -58 | -72.360000 |
| -57 | -70.560000 |
| -56 | -68.760000 |
| -55 | -66.960000 |
| -54 | -65.160000 |
| -53 | -63.360000 |
| -52 | -61.560000 |
| -51 | -59.760000 |
| -50 | -57.960000 |
| -49 | -56.160000 |
| -48 | -54.360000 |
| -47 | -52.560000 |
| -46 | -50.760000 |
| -45 | -48.960000 |
| -44 | -47.160000 |
| -43 | -45.360000 |
| -42 | -43.560000 |
| -41 | -41.760000 |
| -40 | -39.960000 |
| -39 | -38.160000 |
| -38 | -36.360000 |
| -37 | -34.560000 |
| -36 | -32.760000 |
| -35 | -30.960000 |
| -34 | -29.160000 |
| -33 | -27.360000 |
| -32 | -25.560000 |
| -31 | -23.760000 |
| -30 | -21.960000 |
| -29 | -20.160000 |
| -28 | -18.360000 |
| -27 | -16.560000 |
| -26 | -14.760000 |
| -25 | -12.960000 |
| -24 | -11.160000 |
| -23 | -9.360000 |
| -22 | -7.560000 |
| -21 | -5.760000 |
| -20 | -3.960000 |
| -19 | -2.160000 |
| -18 | -.360000 |
| -17 | 1.440000 |
| -16 | 3.240000 |
| -15 | 5.040000 |
| -14 | 6.840000 |
| -13 | 8.640000 |
| -12 | 10.440000 |
| -11 | 12.240000 |
| -10 | 14.040000 |
| -9 | 15.840000 |
| -8 | 17.640000 |
| -7 | 19.440000 |
| -6 | 21.240000 |
| -5 | 23.040000 |
| -4 | 24.840000 |
| -3 | 26.640000 |
| -2 | 28.440000 |
| -1 | 30.240000 |
| 0 | 32.040000 |

## TEMPERATURE
### Celsius to Fahrenheit (cont'd)

| °C | °F | °C | °F |
|---|---|---|---|
| 1 | 33.840000 | 51 | 123.840000 |
| 2 | 35.640000 | 52 | 125.640000 |
| 3 | 37.440000 | 53 | 127.440000 |
| 4 | 39.240000 | 54 | 129.240000 |
| 5 | 41.040000 | 55 | 131.040000 |
| 6 | 42.840000 | 56 | 132.840000 |
| 7 | 44.640000 | 57 | 134.640000 |
| 8 | 46.440000 | 58 | 136.440000 |
| 9 | 48.240000 | 59 | 138.240000 |
| 10 | 50.040000 | 60 | 140.040000 |
| 11 | 51.840000 | 61 | 141.840000 |
| 12 | 53.640000 | 62 | 143.640000 |
| 13 | 55.440000 | 63 | 145.440000 |
| 14 | 57.240000 | 64 | 147.240000 |
| 15 | 59.040000 | 65 | 149.040000 |
| 16 | 60.840000 | 66 | 150.840000 |
| 17 | 62.640000 | 67 | 152.640000 |
| 18 | 64.440000 | 68 | 154.440000 |
| 19 | 66.240000 | 69 | 156.240000 |
| 20 | 68.040000 | 70 | 158.040000 |
| 21 | 69.840000 | 71 | 159.840000 |
| 22 | 71.640000 | 72 | 161.640000 |
| 23 | 73.440000 | 73 | 163.440000 |
| 24 | 75.240000 | 74 | 165.240000 |
| 25 | 77.040000 | 75 | 167.040000 |
| 26 | 78.840000 | 76 | 168.840000 |
| 27 | 80.640000 | 77 | 170.640000 |
| 28 | 82.440000 | 78 | 172.440000 |
| 29 | 84.240000 | 79 | 174.240000 |
| 30 | 86.040000 | 80 | 176.040000 |
| 31 | 87.840000 | 81 | 177.840000 |
| 32 | 89.640000 | 82 | 179.640000 |
| 33 | 91.440000 | 83 | 181.440000 |
| 34 | 93.240000 | 84 | 183.240000 |
| 35 | 95.040000 | 85 | 185.040000 |
| 36 | 96.840000 | 86 | 186.840000 |
| 37 | 98.640000 | 87 | 188.640000 |
| 38 | 100.440000 | 88 | 190.440000 |
| 39 | 102.240000 | 89 | 192.240000 |
| 40 | 104.040000 | 90 | 194.040000 |
| 41 | 105.840000 | 91 | 195.840000 |
| 42 | 107.640000 | 92 | 197.640000 |
| 43 | 109.440000 | 93 | 199.440000 |
| 44 | 111.240000 | 94 | 201.240000 |
| 45 | 113.040000 | 95 | 203.040000 |
| 46 | 114.840000 | 96 | 204.840000 |
| 47 | 116.640000 | 97 | 206.640000 |
| 48 | 118.440000 | 98 | 208.440000 |
| 49 | 120.240000 | 99 | 210.240000 |
| 50 | 122.040000 | 100 | 212.040000 |

TEMPERATURE
Celsius to Fahrenheit (cont'd)

| °C | °F | °C | °F |
|---|---|---|---|
| 105 | 221.040000 | 305 | 581.040000 |
| 110 | 230.040000 | 310 | 590.040000 |
| 115 | 239.040000 | 315 | 599.040000 |
| 120 | 248.040000 | 320 | 608.040000 |
| 125 | 257.040000 | 325 | 617.040000 |
| 130 | 266.040000 | 330 | 626.040000 |
| 135 | 275.040000 | 335 | 635.040000 |
| 140 | 284.040000 | 340 | 644.040000 |
| 145 | 293.040000 | 345 | 653.040000 |
| 150 | 302.040000 | 350 | 662.040000 |
| 155 | 311.040000 | 355 | 671.040000 |
| 160 | 320.040000 | 360 | 680.040000 |
| 165 | 329.040000 | 365 | 689.040000 |
| 170 | 338.040000 | 370 | 698.040000 |
| 175 | 347.040000 | 375 | 707.040000 |
| 180 | 356.040000 | 380 | 716.040000 |
| 185 | 365.040000 | 385 | 725.040000 |
| 190 | 374.040000 | 390 | 734.040000 |
| 195 | 383.040000 | 395 | 743.040000 |
| 200 | 392.040000 | 400 | 752.040000 |
| 205 | 401.040000 | 405 | 761.040000 |
| 210 | 410.040000 | 410 | 770.040000 |
| 215 | 419.040000 | 415 | 779.040000 |
| 220 | 428.040000 | 420 | 788.040000 |
| 225 | 437.040000 | 425 | 797.040000 |
| 230 | 446.040000 | 430 | 806.040000 |
| 235 | 455.040000 | 435 | 815.040000 |
| 240 | 464.040000 | 440 | 824.040000 |
| 245 | 473.040000 | 445 | 833.040000 |
| 250 | 482.040000 | 450 | 842.040000 |
| 255 | 491.040000 | 455 | 851.040000 |
| 260 | 500.040000 | 460 | 860.040000 |
| 265 | 509.040000 | 465 | 869.040000 |
| 270 | 518.040000 | 470 | 878.040000 |
| 275 | 527.040000 | 475 | 887.040000 |
| 280 | 536.040000 | 480 | 896.040000 |
| 285 | 545.040000 | 485 | 905.040000 |
| 290 | 554.040000 | 490 | 914.040000 |
| 295 | 563.040000 | 495 | 923.040000 |
| 300 | 572.040000 | 500 | 932.040000 |

## TEMPERATURE
### Celsius to Fahrenheit (cont'd)

| °C | °F | °C | °F |
|---|---|---|---|
| 510 | 950.040000 | 910 | 1670.040000 |
| 520 | 968.040000 | 920 | 1688.040000 |
| 530 | 986.040000 | 930 | 1706.040000 |
| 540 | 1004.040000 | 940 | 1724.040000 |
| 550 | 1022.040000 | 950 | 1742.040000 |
| 560 | 1040.040000 | 960 | 1760.040000 |
| 570 | 1058.040000 | 970 | 1778.040000 |
| 580 | 1076.040000 | 980 | 1796.040000 |
| 590 | 1094.040000 | 990 | 1814.040000 |
| 600 | 1112.040000 | 1000 | 1832.040000 |
| 610 | 1130.040000 | 1010 | 1850.040000 |
| 620 | 1148.040000 | 1020 | 1868.040000 |
| 630 | 1166.040000 | 1030 | 1886.040000 |
| 640 | 1184.040000 | 1040 | 1904.040000 |
| 650 | 1202.040000 | 1050 | 1922.040000 |
| 660 | 1220.040000 | 1060 | 1940.040000 |
| 670 | 1238.040000 | 1070 | 1958.040000 |
| 680 | 1256.040000 | 1080 | 1976.040000 |
| 690 | 1274.040000 | 1090 | 1994.040000 |
| 700 | 1292.040000 | 1100 | 2012.040000 |
| 710 | 1310.040000 | 1110 | 2030.040000 |
| 720 | 1328.040000 | 1120 | 2048.040000 |
| 730 | 1346.040000 | 1130 | 2066.040000 |
| 740 | 1364.040000 | 1140 | 2084.040000 |
| 750 | 1382.040000 | 1150 | 2102.040000 |
| 760 | 1400.040000 | 1160 | 2120.040000 |
| 770 | 1418.040000 | 1170 | 2138.040000 |
| 780 | 1436.040000 | 1180 | 2156.040000 |
| 790 | 1454.040000 | 1190 | 2174.040000 |
| 800 | 1472.040000 | 1200 | 2192.040000 |
| 810 | 1490.040000 | 1210 | 2210.040000 |
| 820 | 1508.040000 | 1220 | 2228.040000 |
| 830 | 1526.040000 | 1230 | 2246.040000 |
| 840 | 1544.040000 | 1240 | 2264.040000 |
| 850 | 1562.040000 | 1250 | 2282.040000 |
| 860 | 1580.040000 | 1260 | 2300.040000 |
| 870 | 1598.040000 | 1270 | 2318.040000 |
| 880 | 1616.040000 | 1280 | 2336.040000 |
| 890 | 1634.040000 | 1290 | 2354.040000 |
| 900 | 1652.040000 | 1300 | 2372.040000 |

## TEMPERATURE
### Celsius to Fahrenheit (cont'd)

| °C | °F | °C | °F |
|---|---|---|---|
| 1310 | 2390.040000 | 1710 | 3110.040000 |
| 1320 | 2408.040000 | 1720 | 3128.040000 |
| 1330 | 2426.040000 | 1730 | 3146.040000 |
| 1340 | 2444.040000 | 1740 | 3164.040000 |
| 1350 | 2462.040000 | 1750 | 3182.040000 |
| 1360 | 2480.040000 | 1760 | 3200.040000 |
| 1370 | 2498.040000 | 1770 | 3218.040000 |
| 1380 | 2516.040000 | 1780 | 3236.040000 |
| 1390 | 2534.040000 | 1790 | 3254.040000 |
| 1400 | 2552.040000 | 1800 | 3272.040000 |
| 1410 | 2570.040000 | 1810 | 3290.040000 |
| 1420 | 2588.040000 | 1820 | 3308.040000 |
| 1430 | 2606.040000 | 1830 | 3326.040000 |
| 1440 | 2624.040000 | 1840 | 3344.040000 |
| 1450 | 2642.040000 | 1850 | 3362.040000 |
| 1460 | 2660.040000 | 1860 | 3380.040000 |
| 1470 | 2678.040000 | 1870 | 3398.040000 |
| 1480 | 2696.040000 | 1880 | 3416.040000 |
| 1490 | 2714.040000 | 1890 | 3434.040000 |
| 1500 | 2732.040000 | 1900 | 3452.040000 |
| 1510 | 2750.040000 | 1910 | 3470.040000 |
| 1520 | 2768.040000 | 1920 | 3488.040000 |
| 1530 | 2786.040000 | 1930 | 3506.040000 |
| 1540 | 2804.040000 | 1940 | 3524.040000 |
| 1550 | 2822.040000 | 1950 | 3542.040000 |
| 1560 | 2840.040000 | 1960 | 3560.040000 |
| 1570 | 2858.040000 | 1970 | 3578.040000 |
| 1580 | 2876.040000 | 1980 | 3596.040000 |
| 1590 | 2894.040000 | 1990 | 3614.040000 |
| 1600 | 2912.040000 | 2000 | 3632.040000 |
| 1610 | 2930.040000 | | |
| 1620 | 2948.040000 | | |
| 1630 | 2966.040000 | | |
| 1640 | 2984.040000 | | |
| 1650 | 3002.040000 | | |
| 1660 | 3020.040000 | | |
| 1670 | 3038.040000 | | |
| 1680 | 3056.040000 | | |
| 1690 | 3074.040000 | | |
| 1700 | 3092.040000 | | |

# VELOCITY
## Centimeters per Second
## to
## Inches per Second

| cm/s | in/s | cm/s | in/s |
|---|---|---|---|
| 1 | .393701 | 51 | 20.078751 |
| 2 | .787402 | 52 | 20.472452 |
| 3 | 1.181103 | 53 | 20.866153 |
| 4 | 1.574804 | 54 | 21.259854 |
| 5 | 1.968505 | 55 | 21.653555 |
| 6 | 2.362206 | 56 | 22.047256 |
| 7 | 2.755907 | 57 | 22.440957 |
| 8 | 3.149608 | 58 | 22.834658 |
| 9 | 3.543309 | 59 | 23.228359 |
| 10 | 3.937010 | 60 | 23.622060 |
| 11 | 4.330711 | 61 | 24.015761 |
| 12 | 4.724412 | 62 | 24.409462 |
| 13 | 5.118113 | 63 | 24.803163 |
| 14 | 5.511814 | 64 | 25.196864 |
| 15 | 5.905515 | 65 | 25.590565 |
| 16 | 6.299216 | 66 | 25.984266 |
| 17 | 6.692917 | 67 | 26.377967 |
| 18 | 7.086618 | 68 | 26.771668 |
| 19 | 7.480319 | 69 | 27.165369 |
| 20 | 7.874020 | 70 | 27.559070 |
| 21 | 8.267721 | 71 | 27.952771 |
| 22 | 8.661422 | 72 | 28.346472 |
| 23 | 9.055123 | 73 | 28.740173 |
| 24 | 9.448824 | 74 | 29.133874 |
| 25 | 9.842525 | 75 | 29.527575 |
| 26 | 10.236226 | 76 | 29.921276 |
| 27 | 10.629927 | 77 | 30.314977 |
| 28 | 11.023628 | 78 | 30.708678 |
| 29 | 11.417329 | 79 | 31.102379 |
| 30 | 11.811030 | 80 | 31.496080 |
| 31 | 12.204731 | 81 | 31.889781 |
| 32 | 12.598432 | 82 | 32.283482 |
| 33 | 12.992133 | 83 | 32.677183 |
| 34 | 13.385834 | 84 | 33.070884 |
| 35 | 13.779535 | 85 | 33.464585 |
| 36 | 14.173236 | 86 | 33.858286 |
| 37 | 14.566937 | 87 | 34.251987 |
| 38 | 14.960638 | 88 | 34.645688 |
| 39 | 15.354339 | 89 | 35.039389 |
| 40 | 15.748040 | 90 | 35.433090 |
| 41 | 16.141741 | 91 | 35.826791 |
| 42 | 16.535442 | 92 | 36.220492 |
| 43 | 16.929143 | 93 | 36.614193 |
| 44 | 17.322844 | 94 | 37.007894 |
| 45 | 17.716545 | 95 | 37.401595 |
| 46 | 18.110246 | 96 | 37.795296 |
| 47 | 18.503947 | 97 | 38.188997 |
| 48 | 18.897648 | 98 | 38.582698 |
| 49 | 19.291349 | 99 | 38.976399 |
| 50 | 19.685050 | 100 | 39.370100 |

## VELOCITY
## Inches per Second to Centimeters per Second

| in/s | cm/s | in/s | cm/s |
|---|---|---|---|
| 1 | 2.540000 | 51 | 129.540000 |
| 2 | 5.080000 | 52 | 132.080000 |
| 3 | 7.620000 | 53 | 134.620000 |
| 4 | 10.160000 | 54 | 137.160000 |
| 5 | 12.700000 | 55 | 139.700000 |
| 6 | 15.240000 | 56 | 142.240000 |
| 7 | 17.780000 | 57 | 144.780000 |
| 8 | 20.320000 | 58 | 147.320000 |
| 9 | 22.860000 | 59 | 149.860000 |
| 10 | 25.400000 | 60 | 152.400000 |
| 11 | 27.940000 | 61 | 154.940000 |
| 12 | 30.480000 | 62 | 157.480000 |
| 13 | 33.020000 | 63 | 160.020000 |
| 14 | 35.560000 | 64 | 162.560000 |
| 15 | 38.100000 | 65 | 165.100000 |
| 16 | 40.640000 | 66 | 167.640000 |
| 17 | 43.180000 | 67 | 170.180000 |
| 18 | 45.720000 | 68 | 172.720000 |
| 19 | 48.260000 | 69 | 175.260000 |
| 20 | 50.800000 | 70 | 177.800000 |
| 21 | 53.340000 | 71 | 180.340000 |
| 22 | 55.880000 | 72 | 182.880000 |
| 23 | 58.420000 | 73 | 185.420000 |
| 24 | 60.960000 | 74 | 187.960000 |
| 25 | 63.500000 | 75 | 190.500000 |
| 26 | 66.040000 | 76 | 193.040000 |
| 27 | 68.580000 | 77 | 195.580000 |
| 28 | 71.120000 | 78 | 198.120000 |
| 29 | 73.660000 | 79 | 200.660000 |
| 30 | 76.200000 | 80 | 203.200000 |
| 31 | 78.740000 | 81 | 205.740000 |
| 32 | 81.280000 | 82 | 208.280000 |
| 33 | 83.820000 | 83 | 210.820000 |
| 34 | 86.360000 | 84 | 213.360000 |
| 35 | 88.900000 | 85 | 215.900000 |
| 36 | 91.440000 | 86 | 218.440000 |
| 37 | 93.980000 | 87 | 220.980000 |
| 38 | 96.520000 | 88 | 223.520000 |
| 39 | 99.060000 | 89 | 226.060000 |
| 40 | 101.600000 | 90 | 228.600000 |
| 41 | 104.140000 | 91 | 231.140000 |
| 42 | 106.680000 | 92 | 233.680000 |
| 43 | 109.220000 | 93 | 236.220000 |
| 44 | 111.760000 | 94 | 238.760000 |
| 45 | 114.300000 | 95 | 241.300000 |
| 46 | 116.840000 | 96 | 243.840000 |
| 47 | 119.380000 | 97 | 246.380000 |
| 48 | 121.920000 | 98 | 248.920000 |
| 49 | 124.460000 | 99 | 251.460000 |
| 50 | 127.000000 | 100 | 254.000000 |

# VELOCITY
## Meters per Second to Feet per Second

| m/s | ft/s | m/s | ft/s |
|---|---|---|---|
| 1 | 3.280839 | 51 | 167.322789 |
| 2 | 6.561678 | 52 | 170.603628 |
| 3 | 9.842517 | 53 | 173.884467 |
| 4 | 13.123356 | 54 | 177.165306 |
| 5 | 16.404195 | 55 | 180.446145 |
| 6 | 19.685034 | 56 | 183.726984 |
| 7 | 22.965873 | 57 | 187.007823 |
| 8 | 26.246712 | 58 | 190.288662 |
| 9 | 29.527551 | 59 | 193.569501 |
| 10 | 32.808390 | 60 | 196.850340 |
| 11 | 36.089229 | 61 | 200.131179 |
| 12 | 39.370068 | 62 | 203.412018 |
| 13 | 42.650907 | 63 | 206.692857 |
| 14 | 45.931746 | 64 | 209.973696 |
| 15 | 49.212585 | 65 | 213.254535 |
| 16 | 52.493424 | 66 | 216.535374 |
| 17 | 55.774263 | 67 | 219.816213 |
| 18 | 59.055102 | 68 | 223.097052 |
| 19 | 62.335941 | 69 | 226.377891 |
| 20 | 65.616780 | 70 | 229.658730 |
| 21 | 68.897619 | 71 | 232.939569 |
| 22 | 72.178458 | 72 | 236.220408 |
| 23 | 75.459297 | 73 | 239.501247 |
| 24 | 78.740136 | 74 | 242.782086 |
| 25 | 82.020975 | 75 | 246.062925 |
| 26 | 85.301814 | 76 | 249.343764 |
| 27 | 88.582653 | 77 | 252.624603 |
| 28 | 91.863492 | 78 | 255.905442 |
| 29 | 95.144331 | 79 | 259.186281 |
| 30 | 98.425170 | 80 | 262.467120 |
| 31 | 101.706009 | 81 | 265.747959 |
| 32 | 104.986848 | 82 | 269.028798 |
| 33 | 108.267687 | 83 | 272.309637 |
| 34 | 111.548526 | 84 | 275.590476 |
| 35 | 114.829365 | 85 | 278.871315 |
| 36 | 118.110204 | 86 | 282.152154 |
| 37 | 121.391043 | 87 | 285.432993 |
| 38 | 124.671882 | 88 | 288.713832 |
| 39 | 127.952721 | 89 | 291.994671 |
| 40 | 131.233560 | 90 | 295.275510 |
| 41 | 134.514399 | 91 | 298.556349 |
| 42 | 137.795238 | 92 | 301.837188 |
| 43 | 141.076077 | 93 | 305.118027 |
| 44 | 144.356916 | 94 | 308.398866 |
| 45 | 147.637755 | 95 | 311.679705 |
| 46 | 150.918594 | 96 | 314.960544 |
| 47 | 154.199433 | 97 | 318.241383 |
| 48 | 157.480272 | 98 | 321.522222 |
| 49 | 160.761111 | 99 | 324.803061 |
| 50 | 164.041950 | 100 | 328.083900 |

## VELOCITY
## Feet per Second
## to
## Meters per Second

| ft/s | m/s | ft/s | m/s |
|---|---|---|---|
| 1 | .304800 | 51 | 15.544800 |
| 2 | .609600 | 52 | 15.849600 |
| 3 | .914400 | 53 | 16.154400 |
| 4 | 1.219200 | 54 | 16.459200 |
| 5 | 1.524000 | 55 | 16.764000 |
| 6 | 1.828800 | 56 | 17.068800 |
| 7 | 2.133600 | 57 | 17.373600 |
| 8 | 2.438400 | 58 | 17.678400 |
| 9 | 2.743200 | 59 | 17.983200 |
| 10 | 3.048000 | 60 | 18.288000 |
| 11 | 3.352800 | 61 | 18.592800 |
| 12 | 3.657600 | 62 | 18.897600 |
| 13 | 3.962400 | 63 | 19.202400 |
| 14 | 4.267200 | 64 | 19.507200 |
| 15 | 4.572000 | 65 | 19.812000 |
| 16 | 4.876800 | 66 | 20.116800 |
| 17 | 5.181600 | 67 | 20.421600 |
| 18 | 5.486400 | 68 | 20.726400 |
| 19 | 5.791200 | 69 | 21.031200 |
| 20 | 6.096000 | 70 | 21.336000 |
| 21 | 6.400800 | 71 | 21.640800 |
| 22 | 6.705600 | 72 | 21.945600 |
| 23 | 7.010400 | 73 | 22.250400 |
| 24 | 7.315200 | 74 | 22.555200 |
| 25 | 7.620000 | 75 | 22.860000 |
| 26 | 7.924800 | 76 | 23.164800 |
| 27 | 8.229600 | 77 | 23.469600 |
| 28 | 8.534400 | 78 | 23.774400 |
| 29 | 8.839200 | 79 | 24.079200 |
| 30 | 9.144000 | 80 | 24.384000 |
| 31 | 9.448800 | 81 | 24.688800 |
| 32 | 9.753600 | 82 | 24.993600 |
| 33 | 10.058400 | 83 | 25.298400 |
| 34 | 10.363200 | 84 | 25.603200 |
| 35 | 10.668000 | 85 | 25.908000 |
| 36 | 10.972800 | 86 | 26.212800 |
| 37 | 11.277600 | 87 | 26.517600 |
| 38 | 11.582400 | 88 | 26.822400 |
| 39 | 11.887200 | 89 | 27.127200 |
| 40 | 12.192000 | 90 | 27.432000 |
| 41 | 12.496800 | 91 | 27.736800 |
| 42 | 12.801600 | 92 | 28.041600 |
| 43 | 13.106400 | 93 | 28.346400 |
| 44 | 13.411200 | 94 | 28.651200 |
| 45 | 13.716000 | 95 | 28.956000 |
| 46 | 14.020800 | 96 | 29.260800 |
| 47 | 14.325600 | 97 | 29.565600 |
| 48 | 14.630400 | 98 | 29.870400 |
| 49 | 14.935200 | 99 | 30.175200 |
| 50 | 15.240000 | 100 | 30.480000 |

## VELOCITY
## Meters per Second
## to
## Miles per Hour

| m/s | mi/h | m/s | mi/h |
|---|---|---|---|
| 1 | 2.236936 | 51 | 114.083736 |
| 2 | 4.473872 | 52 | 116.320672 |
| 3 | 6.710808 | 53 | 118.557608 |
| 4 | 8.947744 | 54 | 120.794544 |
| 5 | 11.184680 | 55 | 123.031480 |
| 6 | 13.421616 | 56 | 125.268416 |
| 7 | 15.658552 | 57 | 127.505352 |
| 8 | 17.895488 | 58 | 129.742288 |
| 9 | 20.132424 | 59 | 131.979224 |
| 10 | 22.369360 | 60 | 134.216160 |
| 11 | 24.606296 | 61 | 136.453096 |
| 12 | 26.843232 | 62 | 138.690032 |
| 13 | 29.080168 | 63 | 140.926968 |
| 14 | 31.317104 | 64 | 143.163904 |
| 15 | 33.554040 | 65 | 145.400840 |
| 16 | 35.790976 | 66 | 147.637776 |
| 17 | 38.027912 | 67 | 149.874712 |
| 18 | 40.264848 | 68 | 152.111648 |
| 19 | 42.501784 | 69 | 154.348584 |
| 20 | 44.738720 | 70 | 156.585520 |
| 21 | 46.975656 | 71 | 158.822456 |
| 22 | 49.212592 | 72 | 161.059392 |
| 23 | 51.449528 | 73 | 163.296328 |
| 24 | 53.686464 | 74 | 165.533264 |
| 25 | 55.923400 | 75 | 167.770200 |
| 26 | 58.160336 | 76 | 170.007136 |
| 27 | 60.397272 | 77 | 172.244072 |
| 28 | 62.634208 | 78 | 174.481008 |
| 29 | 64.871144 | 79 | 176.717944 |
| 30 | 67.108080 | 80 | 178.954880 |
| 31 | 69.345016 | 81 | 181.191816 |
| 32 | 71.581952 | 82 | 183.428752 |
| 33 | 73.818888 | 83 | 185.665688 |
| 34 | 76.055824 | 84 | 187.902624 |
| 35 | 78.292760 | 85 | 190.139560 |
| 36 | 80.529696 | 86 | 192.376496 |
| 37 | 82.766632 | 87 | 194.613432 |
| 38 | 85.003568 | 88 | 196.850368 |
| 39 | 87.240504 | 89 | 199.087304 |
| 40 | 89.477440 | 90 | 201.324240 |
| 41 | 91.714376 | 91 | 203.561176 |
| 42 | 93.951312 | 92 | 205.798112 |
| 43 | 96.188248 | 93 | 208.035048 |
| 44 | 98.425184 | 94 | 210.271984 |
| 45 | 100.662120 | 95 | 212.508920 |
| 46 | 102.899056 | 96 | 214.745856 |
| 47 | 105.135992 | 97 | 216.982792 |
| 48 | 107.372928 | 98 | 219.219728 |
| 49 | 109.609864 | 99 | 221.456664 |
| 50 | 111.846800 | 100 | 223.693600 |

## VELOCITY
### Miles per Hour
### to
### Meters per Second

| mi/h | m/s | mi/h | m/s |
|---|---|---|---|
| 1 | .447040 | 51 | 22.799040 |
| 2 | .894080 | 52 | 23.246080 |
| 3 | 1.341120 | 53 | 23.693120 |
| 4 | 1.788160 | 54 | 24.140160 |
| 5 | 2.235200 | 55 | 24.587200 |
| 6 | 2.682240 | 56 | 25.034240 |
| 7 | 3.129280 | 57 | 25.481280 |
| 8 | 3.576320 | 58 | 25.928320 |
| 9 | 4.023360 | 59 | 26.375360 |
| 10 | 4.470400 | 60 | 26.822400 |
| 11 | 4.917440 | 61 | 27.269440 |
| 12 | 5.364480 | 62 | 27.716480 |
| 13 | 5.811520 | 63 | 28.163520 |
| 14 | 6.258560 | 64 | 28.610560 |
| 15 | 6.705600 | 65 | 29.057600 |
| 16 | 7.152640 | 66 | 29.504640 |
| 17 | 7.599680 | 67 | 29.951680 |
| 18 | 8.046720 | 68 | 30.398720 |
| 19 | 8.493760 | 69 | 30.845760 |
| 20 | 8.940800 | 70 | 31.292800 |
| 21 | 9.387840 | 71 | 31.739840 |
| 22 | 9.834880 | 72 | 32.186880 |
| 23 | 10.281920 | 73 | 32.633920 |
| 24 | 10.728960 | 74 | 33.080960 |
| 25 | 11.176000 | 75 | 33.528000 |
| 26 | 11.623040 | 76 | 33.975040 |
| 27 | 12.070080 | 77 | 34.422080 |
| 28 | 12.517120 | 78 | 34.869120 |
| 29 | 12.964160 | 79 | 35.316160 |
| 30 | 13.411200 | 80 | 35.763200 |
| 31 | 13.858240 | 81 | 36.210240 |
| 32 | 14.305280 | 82 | 36.657280 |
| 33 | 14.752320 | 83 | 37.104320 |
| 34 | 15.199360 | 84 | 37.551360 |
| 35 | 15.646400 | 85 | 37.998400 |
| 36 | 16.093440 | 86 | 38.445440 |
| 37 | 16.540480 | 87 | 38.892480 |
| 38 | 16.987520 | 88 | 39.339520 |
| 39 | 17.434560 | 89 | 39.786560 |
| 40 | 17.881600 | 90 | 40.233600 |
| 41 | 18.328640 | 91 | 40.680640 |
| 42 | 18.775680 | 92 | 41.127680 |
| 43 | 19.222720 | 93 | 41.574720 |
| 44 | 19.669760 | 94 | 42.021760 |
| 45 | 20.116800 | 95 | 42.468800 |
| 46 | 20.563840 | 96 | 42.915840 |
| 47 | 21.010880 | 97 | 43.362880 |
| 48 | 21.457920 | 98 | 43.809920 |
| 49 | 21.904960 | 99 | 44.256960 |
| 50 | 22.352000 | 100 | 44.704000 |

# VELOCITY
## Kilometers per Hour to Miles per Hour

| km/h | mi/h |
|---|---|
| 1 | .621371 |
| 2 | 1.242742 |
| 3 | 1.864113 |
| 4 | 2.485484 |
| 5 | 3.106855 |
| 6 | 3.728226 |
| 7 | 4.349597 |
| 8 | 4.970968 |
| 9 | 5.592339 |
| 10 | 6.213710 |
| 11 | 6.835081 |
| 12 | 7.456452 |
| 13 | 8.077823 |
| 14 | 8.699194 |
| 15 | 9.320565 |
| 16 | 9.941936 |
| 17 | 10.563307 |
| 18 | 11.184678 |
| 19 | 11.806049 |
| 20 | 12.427420 |
| 21 | 13.048791 |
| 22 | 13.670162 |
| 23 | 14.291533 |
| 24 | 14.912904 |
| 25 | 15.534275 |
| 26 | 16.155646 |
| 27 | 16.777017 |
| 28 | 17.398388 |
| 29 | 18.019759 |
| 30 | 18.641130 |
| 31 | 19.262501 |
| 32 | 19.883872 |
| 33 | 20.505243 |
| 34 | 21.126614 |
| 35 | 21.747985 |
| 36 | 22.369356 |
| 37 | 22.990727 |
| 38 | 23.612098 |
| 39 | 24.233469 |
| 40 | 24.854840 |
| 41 | 25.476211 |
| 42 | 26.097582 |
| 43 | 26.718953 |
| 44 | 27.340324 |
| 45 | 27.961695 |
| 46 | 28.583066 |
| 47 | 29.204437 |
| 48 | 29.825808 |
| 49 | 30.447179 |
| 50 | 31.068550 |
| 51 | 31.689921 |
| 52 | 32.311292 |
| 53 | 32.932663 |
| 54 | 33.554034 |
| 55 | 34.175405 |
| 56 | 34.796776 |
| 57 | 35.418147 |
| 58 | 36.039518 |
| 59 | 36.660889 |
| 60 | 37.282260 |
| 61 | 37.903631 |
| 62 | 38.525002 |
| 63 | 39.146373 |
| 64 | 39.767744 |
| 65 | 40.389115 |
| 66 | 41.010486 |
| 67 | 41.631857 |
| 68 | 42.253228 |
| 69 | 42.874599 |
| 70 | 43.495970 |
| 71 | 44.117341 |
| 72 | 44.738712 |
| 73 | 45.360083 |
| 74 | 45.981454 |
| 75 | 46.602825 |
| 76 | 47.224196 |
| 77 | 47.845567 |
| 78 | 48.466938 |
| 79 | 49.088309 |
| 80 | 49.709680 |
| 81 | 50.331051 |
| 82 | 50.952422 |
| 83 | 51.573793 |
| 84 | 52.195164 |
| 85 | 52.816535 |
| 86 | 53.437906 |
| 87 | 54.059277 |
| 88 | 54.680648 |
| 89 | 55.302019 |
| 90 | 55.923390 |
| 91 | 56.544761 |
| 92 | 57.166132 |
| 93 | 57.787503 |
| 94 | 58.408874 |
| 95 | 59.030245 |
| 96 | 59.651616 |
| 97 | 60.272987 |
| 98 | 60.894358 |
| 99 | 61.515729 |
| 100 | 62.137100 |

## VELOCITY
## Miles per Hour
## to
## Kilometers per Hour

| mi/h | km/h | mi/h | km/h |
|---|---|---|---|
| 1 | 1.609344 | 51 | 82.076544 |
| 2 | 3.218688 | 52 | 83.685888 |
| 3 | 4.828032 | 53 | 85.295232 |
| 4 | 6.437376 | 54 | 86.904576 |
| 5 | 8.046720 | 55 | 88.513920 |
| 6 | 9.656064 | 56 | 90.123264 |
| 7 | 11.265408 | 57 | 91.732608 |
| 8 | 12.874752 | 58 | 93.341952 |
| 9 | 14.484096 | 59 | 94.951296 |
| 10 | 16.093440 | 60 | 96.560640 |
| 11 | 17.702784 | 61 | 98.169984 |
| 12 | 19.312128 | 62 | 99.779328 |
| 13 | 20.921472 | 63 | 101.388672 |
| 14 | 22.530816 | 64 | 102.998016 |
| 15 | 24.140160 | 65 | 104.607360 |
| 16 | 25.749504 | 66 | 106.216704 |
| 17 | 27.358848 | 67 | 107.826048 |
| 18 | 28.968192 | 68 | 109.435392 |
| 19 | 30.577536 | 69 | 111.044736 |
| 20 | 32.186880 | 70 | 112.654080 |
| 21 | 33.796224 | 71 | 114.263424 |
| 22 | 35.405568 | 72 | 115.872768 |
| 23 | 37.014912 | 73 | 117.482112 |
| 24 | 38.624256 | 74 | 119.091456 |
| 25 | 40.233600 | 75 | 120.700800 |
| 26 | 41.842944 | 76 | 122.310144 |
| 27 | 43.452288 | 77 | 123.919488 |
| 28 | 45.061632 | 78 | 125.528832 |
| 29 | 46.670976 | 79 | 127.138176 |
| 30 | 48.280320 | 80 | 128.747520 |
| 31 | 49.889664 | 81 | 130.356864 |
| 32 | 51.499008 | 82 | 131.966208 |
| 33 | 53.108352 | 83 | 133.575552 |
| 34 | 54.717696 | 84 | 135.184896 |
| 35 | 56.327040 | 85 | 136.794240 |
| 36 | 57.936384 | 86 | 138.403584 |
| 37 | 59.545728 | 87 | 140.012928 |
| 38 | 61.155072 | 88 | 141.622272 |
| 39 | 62.764416 | 89 | 143.231616 |
| 40 | 64.373760 | 90 | 144.840960 |
| 41 | 65.983104 | 91 | 146.450304 |
| 42 | 67.592448 | 92 | 148.059648 |
| 43 | 69.201792 | 93 | 149.668992 |
| 44 | 70.811136 | 94 | 151.278336 |
| 45 | 72.420480 | 95 | 152.887680 |
| 46 | 74.029824 | 96 | 154.497024 |
| 47 | 75.639168 | 97 | 156.106368 |
| 48 | 77.248512 | 98 | 157.715712 |
| 49 | 78.857856 | 99 | 159.325056 |
| 50 | 80.467200 | 100 | 160.934400 |

## VELOCITY
### Kilometers per Hour to Knots

| km/h | knots | km/h | knots |
|---|---|---|---|
| 1 | .539957 | 51 | 27.537807 |
| 2 | 1.079914 | 52 | 28.077764 |
| 3 | 1.619871 | 53 | 28.617721 |
| 4 | 2.159828 | 54 | 29.157678 |
| 5 | 2.699785 | 55 | 29.697635 |
| 6 | 3.239742 | 56 | 30.237592 |
| 7 | 3.779699 | 57 | 30.777549 |
| 8 | 4.319656 | 58 | 31.317506 |
| 9 | 4.859613 | 59 | 31.857463 |
| 10 | 5.399570 | 60 | 32.397420 |
| 11 | 5.939527 | 61 | 32.937377 |
| 12 | 6.479484 | 62 | 33.477334 |
| 13 | 7.019441 | 63 | 34.017291 |
| 14 | 7.559398 | 64 | 34.557248 |
| 15 | 8.099355 | 65 | 35.097205 |
| 16 | 8.639312 | 66 | 35.637162 |
| 17 | 9.179269 | 67 | 36.177119 |
| 18 | 9.719226 | 68 | 36.717076 |
| 19 | 10.259183 | 69 | 37.257033 |
| 20 | 10.799140 | 70 | 37.796990 |
| 21 | 11.339097 | 71 | 38.336947 |
| 22 | 11.879054 | 72 | 38.876904 |
| 23 | 12.419011 | 73 | 39.416861 |
| 24 | 12.958968 | 74 | 39.956818 |
| 25 | 13.498925 | 75 | 40.496775 |
| 26 | 14.038882 | 76 | 41.036732 |
| 27 | 14.578839 | 77 | 41.576689 |
| 28 | 15.118796 | 78 | 42.116646 |
| 29 | 15.658753 | 79 | 42.656603 |
| 30 | 16.198710 | 80 | 43.196560 |
| 31 | 16.738667 | 81 | 43.736517 |
| 32 | 17.278624 | 82 | 44.276474 |
| 33 | 17.818581 | 83 | 44.816431 |
| 34 | 18.358538 | 84 | 45.356388 |
| 35 | 18.898495 | 85 | 45.896345 |
| 36 | 19.438452 | 86 | 46.436302 |
| 37 | 19.978409 | 87 | 46.976259 |
| 38 | 20.518366 | 88 | 47.516216 |
| 39 | 21.058323 | 89 | 48.056173 |
| 40 | 21.598280 | 90 | 48.596130 |
| 41 | 22.138237 | 91 | 49.136087 |
| 42 | 22.678194 | 92 | 49.676044 |
| 43 | 23.218151 | 93 | 50.216001 |
| 44 | 23.758108 | 94 | 50.755958 |
| 45 | 24.298065 | 95 | 51.295915 |
| 46 | 24.838022 | 96 | 51.835872 |
| 47 | 25.377979 | 97 | 52.375829 |
| 48 | 25.917936 | 98 | 52.915786 |
| 49 | 26.457893 | 99 | 53.455743 |
| 50 | 26.997850 | 100 | 53.995700 |

## VELOCITY
### Knots to Kilometers per Hour

| knots | km/h | knots | km/h |
|---|---|---|---|
| 1 | 1.852000 | 51 | 94.452000 |
| 2 | 3.704000 | 52 | 96.304000 |
| 3 | 5.556000 | 53 | 98.156000 |
| 4 | 7.408000 | 54 | 100.008000 |
| 5 | 9.260000 | 55 | 101.860000 |
| 6 | 11.112000 | 56 | 103.712000 |
| 7 | 12.964000 | 57 | 105.564000 |
| 8 | 14.816000 | 58 | 107.416000 |
| 9 | 16.668000 | 59 | 109.268000 |
| 10 | 18.520000 | 60 | 111.120000 |
| 11 | 20.372000 | 61 | 112.972000 |
| 12 | 22.224000 | 62 | 114.824000 |
| 13 | 24.076000 | 63 | 116.676000 |
| 14 | 25.928000 | 64 | 118.528000 |
| 15 | 27.780000 | 65 | 120.380000 |
| 16 | 29.632000 | 66 | 122.232000 |
| 17 | 31.484000 | 67 | 124.084000 |
| 18 | 33.336000 | 68 | 125.936000 |
| 19 | 35.188000 | 69 | 127.788000 |
| 20 | 37.040000 | 70 | 129.640000 |
| 21 | 38.892000 | 71 | 131.492000 |
| 22 | 40.744000 | 72 | 133.344000 |
| 23 | 42.596000 | 73 | 135.196000 |
| 24 | 44.448000 | 74 | 137.048000 |
| 25 | 46.300000 | 75 | 138.900000 |
| 26 | 48.152000 | 76 | 140.752000 |
| 27 | 50.004000 | 77 | 142.604000 |
| 28 | 51.856000 | 78 | 144.456000 |
| 29 | 53.708000 | 79 | 146.308000 |
| 30 | 55.560000 | 80 | 148.160000 |
| 31 | 57.412000 | 81 | 150.012000 |
| 32 | 59.264000 | 82 | 151.864000 |
| 33 | 61.116000 | 83 | 153.716000 |
| 34 | 62.968000 | 84 | 155.568000 |
| 35 | 64.820000 | 85 | 157.420000 |
| 36 | 66.672000 | 86 | 159.272000 |
| 37 | 68.524000 | 87 | 161.124000 |
| 38 | 70.376000 | 88 | 162.976000 |
| 39 | 72.228000 | 89 | 164.828000 |
| 40 | 74.080000 | 90 | 166.680000 |
| 41 | 75.932000 | 91 | 168.532000 |
| 42 | 77.784000 | 92 | 170.384000 |
| 43 | 79.636000 | 93 | 172.236000 |
| 44 | 81.488000 | 94 | 174.088000 |
| 45 | 83.340000 | 95 | 175.940000 |
| 46 | 85.192000 | 96 | 177.792000 |
| 47 | 87.044000 | 97 | 179.644000 |
| 48 | 88.896000 | 98 | 181.496000 |
| 49 | 90.748000 | 99 | 183.348000 |
| 50 | 92.600000 | 100 | 185.200000 |

## ACCELERATION
## Inches per Second Squared
## to
## Centimeters per Second Squared

| in/s² | cm/s² | in/s² | cm/s² |
|---|---|---|---|
| 1 | 2.540000 | 51 | 129.540000 |
| 2 | 5.080000 | 52 | 132.080000 |
| 3 | 7.620000 | 53 | 134.620000 |
| 4 | 10.160000 | 54 | 137.160000 |
| 5 | 12.700000 | 55 | 139.700000 |
| 6 | 15.240000 | 56 | 142.240000 |
| 7 | 17.780000 | 57 | 144.780000 |
| 8 | 20.320000 | 58 | 147.320000 |
| 9 | 22.860000 | 59 | 149.860000 |
| 10 | 25.400000 | 60 | 152.400000 |
| 11 | 27.940000 | 61 | 154.940000 |
| 12 | 30.480000 | 62 | 157.480000 |
| 13 | 33.020000 | 63 | 160.020000 |
| 14 | 35.560000 | 64 | 162.560000 |
| 15 | 38.100000 | 65 | 165.100000 |
| 16 | 40.640000 | 66 | 167.640000 |
| 17 | 43.180000 | 67 | 170.180000 |
| 18 | 45.720000 | 68 | 172.720000 |
| 19 | 48.260000 | 69 | 175.260000 |
| 20 | 50.800000 | 70 | 177.800000 |
| 21 | 53.340000 | 71 | 180.340000 |
| 22 | 55.880000 | 72 | 182.880000 |
| 23 | 58.420000 | 73 | 185.420000 |
| 24 | 60.960000 | 74 | 187.960000 |
| 25 | 63.500000 | 75 | 190.500000 |
| 26 | 66.040000 | 76 | 193.040000 |
| 27 | 68.580000 | 77 | 195.580000 |
| 28 | 71.120000 | 78 | 198.120000 |
| 29 | 73.660000 | 79 | 200.660000 |
| 30 | 76.200000 | 80 | 203.200000 |
| 31 | 78.740000 | 81 | 205.740000 |
| 32 | 81.280000 | 82 | 208.280000 |
| 33 | 83.820000 | 83 | 210.820000 |
| 34 | 86.360000 | 84 | 213.360000 |
| 35 | 88.900000 | 85 | 215.900000 |
| 36 | 91.440000 | 86 | 218.440000 |
| 37 | 93.980000 | 87 | 220.980000 |
| 38 | 96.520000 | 88 | 223.520000 |
| 39 | 99.060000 | 89 | 226.060000 |
| 40 | 101.600000 | 90 | 228.600000 |
| 41 | 104.140000 | 91 | 231.140000 |
| 42 | 106.680000 | 92 | 233.680000 |
| 43 | 109.220000 | 93 | 236.220000 |
| 44 | 111.760000 | 94 | 238.760000 |
| 45 | 114.300000 | 95 | 241.300000 |
| 46 | 116.840000 | 96 | 243.840000 |
| 47 | 119.380000 | 97 | 246.380000 |
| 48 | 121.920000 | 98 | 248.920000 |
| 49 | 124.460000 | 99 | 251.460000 |
| 50 | 127.000000 | 100 | 254.000000 |

# ACCELERATION
## Centimeters per Second Squared
## to
## Inches per Second Squared

| cm/s$^2$ | in/s$^2$ | cm/s$^2$ | in/s$^2$ |
|---|---|---|---|
| 1 | .393701 | 51 | 20.078751 |
| 2 | .787402 | 52 | 20.472452 |
| 3 | 1.181103 | 53 | 20.866153 |
| 4 | 1.574804 | 54 | 21.259854 |
| 5 | 1.968505 | 55 | 21.653555 |
| 6 | 2.362206 | 56 | 22.047256 |
| 7 | 2.755907 | 57 | 22.440957 |
| 8 | 3.149608 | 58 | 22.834658 |
| 9 | 3.543309 | 59 | 23.228359 |
| 10 | 3.937010 | 60 | 23.622060 |
| 11 | 4.330711 | 61 | 24.015761 |
| 12 | 4.724412 | 62 | 24.409462 |
| 13 | 5.118113 | 63 | 24.803163 |
| 14 | 5.511814 | 64 | 25.196864 |
| 15 | 5.905515 | 65 | 25.590565 |
| 16 | 6.299216 | 66 | 25.984266 |
| 17 | 6.692917 | 67 | 26.377967 |
| 18 | 7.086618 | 68 | 26.771668 |
| 19 | 7.480319 | 69 | 27.165369 |
| 20 | 7.874020 | 70 | 27.559070 |
| 21 | 8.267721 | 71 | 27.952771 |
| 22 | 8.661422 | 72 | 28.346472 |
| 23 | 9.055123 | 73 | 28.740173 |
| 24 | 9.448824 | 74 | 29.133874 |
| 25 | 9.842525 | 75 | 29.527575 |
| 26 | 10.236226 | 76 | 29.921276 |
| 27 | 10.629927 | 77 | 30.314977 |
| 28 | 11.023628 | 78 | 30.708678 |
| 29 | 11.417329 | 79 | 31.102379 |
| 30 | 11.811030 | 80 | 31.496080 |
| 31 | 12.204731 | 81 | 31.889781 |
| 32 | 12.598432 | 82 | 32.283482 |
| 33 | 12.992133 | 83 | 32.677183 |
| 34 | 13.385834 | 84 | 33.070884 |
| 35 | 13.779535 | 85 | 33.464585 |
| 36 | 14.173236 | 86 | 33.858286 |
| 37 | 14.566937 | 87 | 34.251987 |
| 38 | 14.960638 | 88 | 34.645688 |
| 39 | 15.354339 | 89 | 35.039389 |
| 40 | 15.748040 | 90 | 35.433090 |
| 41 | 16.141741 | 91 | 35.826791 |
| 42 | 16.535442 | 92 | 36.220492 |
| 43 | 16.929143 | 93 | 36.614193 |
| 44 | 17.322844 | 94 | 37.007894 |
| 45 | 17.716545 | 95 | 37.401595 |
| 46 | 18.110246 | 96 | 37.795296 |
| 47 | 18.503947 | 97 | 38.188997 |
| 48 | 18.897648 | 98 | 38.582698 |
| 49 | 19.291349 | 99 | 38.976399 |
| 50 | 19.685050 | 100 | 39.370100 |

## ACCELERATION
## Inches per Second Squared
## to
## Meters per Minute Squared

| in/s² | m/min² | in/s² | m/min² |
|---|---|---|---|
| 1 | 91.440000 | 51 | 4663.440000 |
| 2 | 182.880000 | 52 | 4754.880000 |
| 3 | 274.320000 | 53 | 4846.320000 |
| 4 | 365.760000 | 54 | 4937.760000 |
| 5 | 457.200000 | 55 | 5029.200000 |
| 6 | 548.640000 | 56 | 5120.640000 |
| 7 | 640.080000 | 57 | 5212.080000 |
| 8 | 731.520000 | 58 | 5303.520000 |
| 9 | 822.960000 | 59 | 5394.960000 |
| 10 | 914.400000 | 60 | 5486.400000 |
| 11 | 1005.840000 | 61 | 5577.840000 |
| 12 | 1097.280000 | 62 | 5669.280000 |
| 13 | 1188.720000 | 63 | 5760.720000 |
| 14 | 1280.160000 | 64 | 5852.160000 |
| 15 | 1371.600000 | 65 | 5943.600000 |
| 16 | 1463.040000 | 66 | 6035.040000 |
| 17 | 1554.480000 | 67 | 6126.480000 |
| 18 | 1645.920000 | 68 | 6217.920000 |
| 19 | 1737.360000 | 69 | 6309.360000 |
| 20 | 1828.800000 | 70 | 6400.800000 |
| 21 | 1920.240000 | 71 | 6492.240000 |
| 22 | 2011.680000 | 72 | 6583.680000 |
| 23 | 2103.120000 | 73 | 6675.120000 |
| 24 | 2194.560000 | 74 | 6766.560000 |
| 25 | 2286.000000 | 75 | 6858.000000 |
| 26 | 2377.440000 | 76 | 6949.440000 |
| 27 | 2468.880000 | 77 | 7040.880000 |
| 28 | 2560.320000 | 78 | 7132.320000 |
| 29 | 2651.760000 | 79 | 7223.760000 |
| 30 | 2743.200000 | 80 | 7315.200000 |
| 31 | 2834.640000 | 81 | 7406.640000 |
| 32 | 2926.080000 | 82 | 7498.080000 |
| 33 | 3017.520000 | 83 | 7589.520000 |
| 34 | 3108.960000 | 84 | 7680.960000 |
| 35 | 3200.400000 | 85 | 7772.400000 |
| 36 | 3291.840000 | 86 | 7863.840000 |
| 37 | 3383.280000 | 87 | 7955.280000 |
| 38 | 3474.720000 | 88 | 8046.720000 |
| 39 | 3566.160000 | 89 | 8138.160000 |
| 40 | 3657.600000 | 90 | 8229.600000 |
| 41 | 3749.040000 | 91 | 8321.040000 |
| 42 | 3840.480000 | 92 | 8412.480000 |
| 43 | 3931.920000 | 93 | 8503.920000 |
| 44 | 4023.360000 | 94 | 8595.360000 |
| 45 | 4114.800000 | 95 | 8686.800000 |
| 46 | 4206.240000 | 96 | 8778.240000 |
| 47 | 4297.680000 | 97 | 8869.680000 |
| 48 | 4389.120000 | 98 | 8961.120000 |
| 49 | 4480.560000 | 99 | 9052.560000 |
| 50 | 4572.000000 | 100 | 9144.000000 |

# ACCELERATION
## Meters per Minute Squared
## to
## Inches per Second Squared

| m/min² | in/s² | m/min² | in/s² |
|---|---|---|---|
| 1 | .010936 | 51 | .557736 |
| 2 | .021872 | 52 | .568672 |
| 3 | .032808 | 53 | .579608 |
| 4 | .043744 | 54 | .590544 |
| 5 | .054680 | 55 | .601480 |
| 6 | .065616 | 56 | .612416 |
| 7 | .076552 | 57 | .623352 |
| 8 | .087488 | 58 | .634288 |
| 9 | .098424 | 59 | .645224 |
| 10 | .109360 | 60 | .656160 |
| 11 | .120296 | 61 | .667096 |
| 12 | .131232 | 62 | .678032 |
| 13 | .142168 | 63 | .688968 |
| 14 | .153104 | 64 | .699904 |
| 15 | .164040 | 65 | .710840 |
| 16 | .174976 | 66 | .721776 |
| 17 | .185912 | 67 | .732712 |
| 18 | .196848 | 68 | .743648 |
| 19 | .207784 | 69 | .754584 |
| 20 | .218720 | 70 | .765520 |
| 21 | .229656 | 71 | .776456 |
| 22 | .240592 | 72 | .787392 |
| 23 | .251528 | 73 | .798328 |
| 24 | .262464 | 74 | .809264 |
| 25 | .273400 | 75 | .820200 |
| 26 | .284336 | 76 | .831136 |
| 27 | .295272 | 77 | .842072 |
| 28 | .306208 | 78 | .853008 |
| 29 | .317144 | 79 | .863944 |
| 30 | .328080 | 80 | .874880 |
| 31 | .339016 | 81 | .885816 |
| 32 | .349952 | 82 | .896752 |
| 33 | .360888 | 83 | .907688 |
| 34 | .371824 | 84 | .918624 |
| 35 | .382760 | 85 | .929560 |
| 36 | .393696 | 86 | .940496 |
| 37 | .404632 | 87 | .951432 |
| 38 | .415568 | 88 | .962368 |
| 39 | .426504 | 89 | .973304 |
| 40 | .437440 | 90 | .984240 |
| 41 | .448376 | 91 | .995176 |
| 42 | .459312 | 92 | 1.006112 |
| 43 | .470248 | 93 | 1.017048 |
| 44 | .481184 | 94 | 1.027984 |
| 45 | .492120 | 95 | 1.038920 |
| 46 | .503056 | 96 | 1.049856 |
| 47 | .513992 | 97 | 1.060792 |
| 48 | .524928 | 98 | 1.071728 |
| 49 | .535864 | 99 | 1.082664 |
| 50 | .546800 | 100 | 1.093600 |

## ACCELERATION
### Feet per Second Squared
### to
### Centimeters per Second Squared

| ft/s² | cm/s² | ft/s² | cm/s² |
|---|---|---|---|
| 1 | 30.480000 | 51 | 1554.480000 |
| 2 | 60.960000 | 52 | 1584.960000 |
| 3 | 91.440000 | 53 | 1615.440000 |
| 4 | 121.920000 | 54 | 1645.920000 |
| 5 | 152.400000 | 55 | 1676.400000 |
| 6 | 182.880000 | 56 | 1706.880000 |
| 7 | 213.360000 | 57 | 1737.360000 |
| 8 | 243.840000 | 58 | 1767.840000 |
| 9 | 274.320000 | 59 | 1798.320000 |
| 10 | 304.800000 | 60 | 1828.800000 |
| 11 | 335.280000 | 61 | 1859.280000 |
| 12 | 365.760000 | 62 | 1889.760000 |
| 13 | 396.240000 | 63 | 1920.240000 |
| 14 | 426.720000 | 64 | 1950.720000 |
| 15 | 457.200000 | 65 | 1981.200000 |
| 16 | 487.680000 | 66 | 2011.680000 |
| 17 | 518.160000 | 67 | 2042.160000 |
| 18 | 548.640000 | 68 | 2072.640000 |
| 19 | 579.120000 | 69 | 2103.120000 |
| 20 | 609.600000 | 70 | 2133.600000 |
| 21 | 640.080000 | 71 | 2164.080000 |
| 22 | 670.560000 | 72 | 2194.560000 |
| 23 | 701.040000 | 73 | 2225.040000 |
| 24 | 731.520000 | 74 | 2255.520000 |
| 25 | 762.000000 | 75 | 2286.000000 |
| 26 | 792.480000 | 76 | 2316.480000 |
| 27 | 822.960000 | 77 | 2346.960000 |
| 28 | 853.440000 | 78 | 2377.440000 |
| 29 | 883.920000 | 79 | 2407.920000 |
| 30 | 914.400000 | 80 | 2438.400000 |
| 31 | 944.880000 | 81 | 2468.880000 |
| 32 | 975.360000 | 82 | 2499.360000 |
| 33 | 1005.840000 | 83 | 2529.840000 |
| 34 | 1036.320000 | 84 | 2560.320000 |
| 35 | 1066.800000 | 85 | 2590.800000 |
| 36 | 1097.280000 | 86 | 2621.280000 |
| 37 | 1127.760000 | 87 | 2651.760000 |
| 38 | 1158.240000 | 88 | 2682.240000 |
| 39 | 1188.720000 | 89 | 2712.720000 |
| 40 | 1219.200000 | 90 | 2743.200000 |
| 41 | 1249.680000 | 91 | 2773.680000 |
| 42 | 1280.160000 | 92 | 2804.160000 |
| 43 | 1310.640000 | 93 | 2834.640000 |
| 44 | 1341.120000 | 94 | 2865.120000 |
| 45 | 1371.600000 | 95 | 2895.600000 |
| 46 | 1402.080000 | 96 | 2926.080000 |
| 47 | 1432.560000 | 97 | 2956.560000 |
| 48 | 1463.040000 | 98 | 2987.040000 |
| 49 | 1493.520000 | 99 | 3017.520000 |
| 50 | 1524.000000 | 100 | 3048.000000 |

## ACCELERATION
## Centimeters per Second Squared to Feet per Second Squared

| cm/s² | ft/s² | cm/s² | ft/s² |
|---|---|---|---|
| 1 | .032808 | 51 | 1.673208 |
| 2 | .065616 | 52 | 1.706016 |
| 3 | .098424 | 53 | 1.738824 |
| 4 | .131232 | 54 | 1.771632 |
| 5 | .164040 | 55 | 1.804440 |
| 6 | .196848 | 56 | 1.837248 |
| 7 | .229656 | 57 | 1.870056 |
| 8 | .262464 | 58 | 1.902864 |
| 9 | .295272 | 59 | 1.935672 |
| 10 | .328080 | 60 | 1.968480 |
| 11 | .360888 | 61 | 2.001288 |
| 12 | .393696 | 62 | 2.034096 |
| 13 | .426504 | 63 | 2.066904 |
| 14 | .459312 | 64 | 2.099712 |
| 15 | .492120 | 65 | 2.132520 |
| 16 | .524928 | 66 | 2.165328 |
| 17 | .557736 | 67 | 2.198136 |
| 18 | .590544 | 68 | 2.230944 |
| 19 | .623352 | 69 | 2.263752 |
| 20 | .656160 | 70 | 2.296560 |
| 21 | .688968 | 71 | 2.329368 |
| 22 | .721776 | 72 | 2.362176 |
| 23 | .754584 | 73 | 2.394984 |
| 24 | .787392 | 74 | 2.427792 |
| 25 | .820200 | 75 | 2.460600 |
| 26 | .853008 | 76 | 2.493408 |
| 27 | .885816 | 77 | 2.526216 |
| 28 | .918624 | 78 | 2.559024 |
| 29 | .951432 | 79 | 2.591832 |
| 30 | .984240 | 80 | 2.624640 |
| 31 | 1.017048 | 81 | 2.657448 |
| 32 | 1.049856 | 82 | 2.690256 |
| 33 | 1.082664 | 83 | 2.723064 |
| 34 | 1.115472 | 84 | 2.755872 |
| 35 | 1.148280 | 85 | 2.788680 |
| 36 | 1.181088 | 86 | 2.821488 |
| 37 | 1.213896 | 87 | 2.854296 |
| 38 | 1.246704 | 88 | 2.887104 |
| 39 | 1.279512 | 89 | 2.919912 |
| 40 | 1.312320 | 90 | 2.952720 |
| 41 | 1.345128 | 91 | 2.985528 |
| 42 | 1.377936 | 92 | 3.018336 |
| 43 | 1.410744 | 93 | 3.051144 |
| 44 | 1.443552 | 94 | 3.083952 |
| 45 | 1.476360 | 95 | 3.116760 |
| 46 | 1.509168 | 96 | 3.149568 |
| 47 | 1.541976 | 97 | 3.182376 |
| 48 | 1.574784 | 98 | 3.215184 |
| 49 | 1.607592 | 99 | 3.247992 |
| 50 | 1.640400 | 100 | 3.280800 |

## ACCELERATION
## Feet per Second Squared
## to
## Meters per Minute Squared

| $ft/s^2$ | $m/min^2$ | $ft/s^2$ | $m/min^2$ |
|---|---|---|---|
| 1 | 1097.280000 | 51 | 55961.280000 |
| 2 | 2194.560000 | 52 | 57058.560000 |
| 3 | 3291.840000 | 53 | 58155.840000 |
| 4 | 4389.120000 | 54 | 59253.120000 |
| 5 | 5486.400000 | 55 | 60350.400000 |
| 6 | 6583.680000 | 56 | 61447.680000 |
| 7 | 7680.960000 | 57 | 62544.960000 |
| 8 | 8778.240000 | 58 | 63642.240000 |
| 9 | 9875.520000 | 59 | 64739.520000 |
| 10 | 10972.800000 | 60 | 65836.800000 |
| 11 | 12070.080000 | 61 | 66934.080000 |
| 12 | 13167.360000 | 62 | 68031.360000 |
| 13 | 14264.640000 | 63 | 69128.640000 |
| 14 | 15361.920000 | 64 | 70225.920000 |
| 15 | 16459.200000 | 65 | 71323.200000 |
| 16 | 17556.480000 | 66 | 72420.480000 |
| 17 | 18653.760000 | 67 | 73517.760000 |
| 18 | 19751.040000 | 68 | 74615.040000 |
| 19 | 20848.320000 | 69 | 75712.320000 |
| 20 | 21945.600000 | 70 | 76809.600000 |
| 21 | 23042.880000 | 71 | 77906.880000 |
| 22 | 24140.160000 | 72 | 79004.160000 |
| 23 | 25237.440000 | 73 | 80101.440000 |
| 24 | 26334.720000 | 74 | 81198.720000 |
| 25 | 27432.000000 | 75 | 82296.000000 |
| 26 | 28529.280000 | 76 | 83393.280000 |
| 27 | 29626.560000 | 77 | 84490.560000 |
| 28 | 30723.840000 | 78 | 85587.840000 |
| 29 | 31821.120000 | 79 | 86685.120000 |
| 30 | 32918.400000 | 80 | 87782.400000 |
| 31 | 34015.680000 | 81 | 88879.680000 |
| 32 | 35112.960000 | 82 | 89976.960000 |
| 33 | 36210.240000 | 83 | 91074.240000 |
| 34 | 37307.520000 | 84 | 92171.520000 |
| 35 | 38404.800000 | 85 | 93268.800000 |
| 36 | 39502.080000 | 86 | 94366.080000 |
| 37 | 40599.360000 | 87 | 95463.360000 |
| 38 | 41696.640000 | 88 | 96560.640000 |
| 39 | 42793.920000 | 89 | 97657.920000 |
| 40 | 43891.200000 | 90 | 98755.200000 |
| 41 | 44988.480000 | 91 | 99852.480000 |
| 42 | 46085.760000 | 92 | 100949.760000 |
| 43 | 47183.040000 | 93 | 102047.040000 |
| 44 | 48280.320000 | 94 | 103144.320000 |
| 45 | 49377.600000 | 95 | 104241.600000 |
| 46 | 50474.880000 | 96 | 105338.880000 |
| 47 | 51572.160000 | 97 | 106436.160000 |
| 48 | 52669.440000 | 98 | 107533.440000 |
| 49 | 53766.720000 | 99 | 108630.720000 |
| 50 | 54864.000000 | 100 | 109728.000000 |

## ACCELERATION
## Meters per Minute Squared to Feet per Second Squared

| m/min² | ft/s² | m/min² | ft/s² |
|---|---|---|---|
| 1 | .000911 | 51 | .046461 |
| 2 | .001822 | 52 | .047372 |
| 3 | .002733 | 53 | .048283 |
| 4 | .003644 | 54 | .049194 |
| 5 | .004555 | 55 | .050105 |
| 6 | .005466 | 56 | .051016 |
| 7 | .006377 | 57 | .051927 |
| 8 | .007288 | 58 | .052838 |
| 9 | .008199 | 59 | .053749 |
| 10 | .009110 | 60 | .054660 |
| 11 | .010021 | 61 | .055571 |
| 12 | .010932 | 62 | .056482 |
| 13 | .011843 | 63 | .057393 |
| 14 | .012754 | 64 | .058304 |
| 15 | .013665 | 65 | .059215 |
| 16 | .014576 | 66 | .060126 |
| 17 | .015487 | 67 | .061037 |
| 18 | .016398 | 68 | .061948 |
| 19 | .017309 | 69 | .062859 |
| 20 | .018220 | 70 | .063770 |
| 21 | .019131 | 71 | .064681 |
| 22 | .020042 | 72 | .065592 |
| 23 | .020953 | 73 | .066503 |
| 24 | .021864 | 74 | .067414 |
| 25 | .022775 | 75 | .068325 |
| 26 | .023686 | 76 | .069236 |
| 27 | .024597 | 77 | .070147 |
| 28 | .025508 | 78 | .071058 |
| 29 | .026419 | 79 | .071969 |
| 30 | .027330 | 80 | .072880 |
| 31 | .028241 | 81 | .073791 |
| 32 | .029152 | 82 | .074702 |
| 33 | .030063 | 83 | .075613 |
| 34 | .030974 | 84 | .076524 |
| 35 | .031885 | 85 | .077435 |
| 36 | .032796 | 86 | .078346 |
| 37 | .033707 | 87 | .079257 |
| 38 | .034618 | 88 | .080168 |
| 39 | .035529 | 89 | .081079 |
| 40 | .036440 | 90 | .081990 |
| 41 | .037351 | 91 | .082901 |
| 42 | .038262 | 92 | .083812 |
| 43 | .039173 | 93 | .084723 |
| 44 | .040084 | 94 | .085634 |
| 45 | .040995 | 95 | .086545 |
| 46 | .041906 | 96 | .087456 |
| 47 | .042817 | 97 | .088367 |
| 48 | .043728 | 98 | .089278 |
| 49 | .044639 | 99 | .090189 |
| 50 | .045550 | 100 | .091100 |

# ACCELERATION
## Feet per Minute Squared
## to
## Centimeters per Second Squared

| ft/min² | cm/s² | ft/min² | cm/s² |
|---|---|---|---|
| 1 | .008467 | 51 | .431817 |
| 2 | .016934 | 52 | .440284 |
| 3 | .025401 | 53 | .448751 |
| 4 | .033868 | 54 | .457218 |
| 5 | .042335 | 55 | .465685 |
| 6 | .050802 | 56 | .474152 |
| 7 | .059269 | 57 | .482619 |
| 8 | .067736 | 58 | .491086 |
| 9 | .076203 | 59 | .499553 |
| 10 | .084670 | 60 | .508020 |
| 11 | .093137 | 61 | .516487 |
| 12 | .101604 | 62 | .524954 |
| 13 | .110071 | 63 | .533421 |
| 14 | .118538 | 64 | .541888 |
| 15 | .127005 | 65 | .550355 |
| 16 | .135472 | 66 | .558822 |
| 17 | .143939 | 67 | .567289 |
| 18 | .152406 | 68 | .575756 |
| 19 | .160873 | 69 | .584223 |
| 20 | .169340 | 70 | .592690 |
| 21 | .177807 | 71 | .601157 |
| 22 | .186274 | 72 | .609624 |
| 23 | .194741 | 73 | .618091 |
| 24 | .203208 | 74 | .626558 |
| 25 | .211675 | 75 | .635025 |
| 26 | .220142 | 76 | .643492 |
| 27 | .228609 | 77 | .651959 |
| 28 | .237076 | 78 | .660426 |
| 29 | .245543 | 79 | .668893 |
| 30 | .254010 | 80 | .677360 |
| 31 | .262477 | 81 | .685827 |
| 32 | .270944 | 82 | .694294 |
| 33 | .279411 | 83 | .702761 |
| 34 | .287878 | 84 | .711228 |
| 35 | .296345 | 85 | .719695 |
| 36 | .304812 | 86 | .728162 |
| 37 | .313279 | 87 | .736629 |
| 38 | .321746 | 88 | .745096 |
| 39 | .330213 | 89 | .753563 |
| 40 | .338680 | 90 | .762030 |
| 41 | .347147 | 91 | .770497 |
| 42 | .355614 | 92 | .778964 |
| 43 | .364081 | 93 | .787431 |
| 44 | .372548 | 94 | .795898 |
| 45 | .381015 | 95 | .804365 |
| 46 | .389482 | 96 | .812832 |
| 47 | .397949 | 97 | .821299 |
| 48 | .406416 | 98 | .829766 |
| 49 | .414883 | 99 | .838233 |
| 50 | .423350 | 100 | .846700 |

## ACCELERATION
## Centimeters per Second Squared
## to
## Feet per Minute Squared

| cm/s² | ft/min² | cm/s² | ft/min² |
|---|---|---|---|
| 1 | 118.110000 | 51 | 6023.610000 |
| 2 | 236.220000 | 52 | 6141.720000 |
| 3 | 354.330000 | 53 | 6259.830000 |
| 4 | 472.440000 | 54 | 6377.940000 |
| 5 | 590.550000 | 55 | 6496.050000 |
| 6 | 708.660000 | 56 | 6614.160000 |
| 7 | 826.770000 | 57 | 6732.270000 |
| 8 | 944.880000 | 58 | 6850.380000 |
| 9 | 1062.990000 | 59 | 6968.490000 |
| 10 | 1181.100000 | 60 | 7086.600000 |
| 11 | 1299.210000 | 61 | 7204.710000 |
| 12 | 1417.320000 | 62 | 7322.820000 |
| 13 | 1535.430000 | 63 | 7440.930000 |
| 14 | 1653.540000 | 64 | 7559.040000 |
| 15 | 1771.650000 | 65 | 7677.150000 |
| 16 | 1889.760000 | 66 | 7795.260000 |
| 17 | 2007.870000 | 67 | 7913.370000 |
| 18 | 2125.980000 | 68 | 8031.480000 |
| 19 | 2244.090000 | 69 | 8149.590000 |
| 20 | 2362.200000 | 70 | 8267.700000 |
| 21 | 2480.310000 | 71 | 8385.810000 |
| 22 | 2598.420000 | 72 | 8503.920000 |
| 23 | 2716.530000 | 73 | 8622.030000 |
| 24 | 2834.640000 | 74 | 8740.140000 |
| 25 | 2952.750000 | 75 | 8858.250000 |
| 26 | 3070.860000 | 76 | 8976.360000 |
| 27 | 3188.970000 | 77 | 9094.470000 |
| 28 | 3307.080000 | 78 | 9212.580000 |
| 29 | 3425.190000 | 79 | 9330.690000 |
| 30 | 3543.300000 | 80 | 9448.800000 |
| 31 | 3661.410000 | 81 | 9566.910000 |
| 32 | 3779.520000 | 82 | 9685.020000 |
| 33 | 3897.630000 | 83 | 9803.130000 |
| 34 | 4015.740000 | 84 | 9921.240000 |
| 35 | 4133.850000 | 85 | 10039.350000 |
| 36 | 4251.960000 | 86 | 10157.460000 |
| 37 | 4370.070000 | 87 | 10275.570000 |
| 38 | 4488.180000 | 88 | 10393.680000 |
| 39 | 4606.290000 | 89 | 10511.790000 |
| 40 | 4724.400000 | 90 | 10629.900000 |
| 41 | 4842.510000 | 91 | 10748.010000 |
| 42 | 4960.620000 | 92 | 10866.120000 |
| 43 | 5078.730000 | 93 | 10984.230000 |
| 44 | 5196.840000 | 94 | 11102.340000 |
| 45 | 5314.950000 | 95 | 11220.450000 |
| 46 | 5433.060000 | 96 | 11338.560000 |
| 47 | 5551.170000 | 97 | 11456.670000 |
| 48 | 5669.280000 | 98 | 11574.780000 |
| 49 | 5787.390000 | 99 | 11692.890000 |
| 50 | 5905.500000 | 100 | 11811.000000 |

## ACCELERATION
## Feet per Second Squared
## to
## Meters per Second Squared

| $ft/s^2$ | $m/s^2$ | $ft/s^2$ | $m/s^2$ |
|---|---|---|---|
| 1 | .304800 | 51 | 15.544800 |
| 2 | .609600 | 52 | 15.849600 |
| 3 | .914400 | 53 | 16.154400 |
| 4 | 1.219200 | 54 | 16.459200 |
| 5 | 1.524000 | 55 | 16.764000 |
| 6 | 1.828800 | 56 | 17.068800 |
| 7 | 2.133600 | 57 | 17.373600 |
| 8 | 2.438400 | 58 | 17.678400 |
| 9 | 2.743200 | 59 | 17.983200 |
| 10 | 3.048000 | 60 | 18.288000 |
| 11 | 3.352800 | 61 | 18.592800 |
| 12 | 3.657600 | 62 | 18.897600 |
| 13 | 3.962400 | 63 | 19.202400 |
| 14 | 4.267200 | 64 | 19.507200 |
| 15 | 4.572000 | 65 | 19.812000 |
| 16 | 4.876800 | 66 | 20.116800 |
| 17 | 5.181600 | 67 | 20.421600 |
| 18 | 5.486400 | 68 | 20.726400 |
| 19 | 5.791200 | 69 | 21.031200 |
| 20 | 6.096000 | 70 | 21.336000 |
| 21 | 6.400800 | 71 | 21.640800 |
| 22 | 6.705600 | 72 | 21.945600 |
| 23 | 7.010400 | 73 | 22.250400 |
| 24 | 7.315200 | 74 | 22.555200 |
| 25 | 7.620000 | 75 | 22.860000 |
| 26 | 7.924800 | 76 | 23.164800 |
| 27 | 8.229600 | 77 | 23.469600 |
| 28 | 8.534400 | 78 | 23.774400 |
| 29 | 8.839200 | 79 | 24.079200 |
| 30 | 9.144000 | 80 | 24.384000 |
| 31 | 9.448800 | 81 | 24.688800 |
| 32 | 9.753600 | 82 | 24.993600 |
| 33 | 10.058400 | 83 | 25.298400 |
| 34 | 10.363200 | 84 | 25.603200 |
| 35 | 10.668000 | 85 | 25.908000 |
| 36 | 10.972800 | 86 | 26.212800 |
| 37 | 11.277600 | 87 | 26.517600 |
| 38 | 11.582400 | 88 | 26.822400 |
| 39 | 11.887200 | 89 | 27.127200 |
| 40 | 12.192000 | 90 | 27.432000 |
| 41 | 12.496800 | 91 | 27.736800 |
| 42 | 12.801600 | 92 | 28.041600 |
| 43 | 13.106400 | 93 | 28.346400 |
| 44 | 13.411200 | 94 | 28.651200 |
| 45 | 13.716000 | 95 | 28.956000 |
| 46 | 14.020800 | 96 | 29.260800 |
| 47 | 14.325600 | 97 | 29.565600 |
| 48 | 14.630400 | 98 | 29.870400 |
| 49 | 14.935200 | 99 | 30.175200 |
| 50 | 15.240000 | 100 | 30.480000 |

## ACCELERATION
## Meters per Second Squared to Feet per Second Squared

| m/s² | ft/s² | m/s² | ft/s² |
|---|---|---|---|
| 1 | 3.280840 | 51 | 167.322840 |
| 2 | 6.561680 | 52 | 170.603680 |
| 3 | 9.842520 | 53 | 173.884520 |
| 4 | 13.123360 | 54 | 177.165360 |
| 5 | 16.404200 | 55 | 180.446200 |
| 6 | 19.685040 | 56 | 183.727040 |
| 7 | 22.965880 | 57 | 187.007880 |
| 8 | 26.246720 | 58 | 190.288720 |
| 9 | 29.527560 | 59 | 193.569560 |
| 10 | 32.808400 | 60 | 196.850400 |
| 11 | 36.089240 | 61 | 200.131240 |
| 12 | 39.370080 | 62 | 203.412080 |
| 13 | 42.650920 | 63 | 206.692920 |
| 14 | 45.931760 | 64 | 209.973760 |
| 15 | 49.212600 | 65 | 213.254600 |
| 16 | 52.493440 | 66 | 216.535440 |
| 17 | 55.774280 | 67 | 219.816280 |
| 18 | 59.055120 | 68 | 223.097120 |
| 19 | 62.335960 | 69 | 226.377960 |
| 20 | 65.616800 | 70 | 229.658800 |
| 21 | 68.897640 | 71 | 232.939640 |
| 22 | 72.178480 | 72 | 236.220480 |
| 23 | 75.459320 | 73 | 239.501320 |
| 24 | 78.740160 | 74 | 242.782160 |
| 25 | 82.021000 | 75 | 246.063000 |
| 26 | 85.301840 | 76 | 249.343840 |
| 27 | 88.582680 | 77 | 252.624680 |
| 28 | 91.863520 | 78 | 255.905520 |
| 29 | 95.144360 | 79 | 259.186360 |
| 30 | 98.425200 | 80 | 262.467200 |
| 31 | 101.706040 | 81 | 265.748040 |
| 32 | 104.986880 | 82 | 269.028880 |
| 33 | 108.267720 | 83 | 272.309720 |
| 34 | 111.548560 | 84 | 275.590560 |
| 35 | 114.829400 | 85 | 278.871400 |
| 36 | 118.110240 | 86 | 282.152240 |
| 37 | 121.391080 | 87 | 285.433080 |
| 38 | 124.671920 | 88 | 288.713920 |
| 39 | 127.952760 | 89 | 291.994760 |
| 40 | 131.233600 | 90 | 295.275600 |
| 41 | 134.514440 | 91 | 298.556440 |
| 42 | 137.795280 | 92 | 301.837280 |
| 43 | 141.076120 | 93 | 305.118120 |
| 44 | 144.356960 | 94 | 308.398960 |
| 45 | 147.637800 | 95 | 311.679800 |
| 46 | 150.918640 | 96 | 314.960640 |
| 47 | 154.199480 | 97 | 318.241480 |
| 48 | 157.480320 | 98 | 321.522320 |
| 49 | 160.761160 | 99 | 324.803160 |
| 50 | 164.042000 | 100 | 328.084000 |

## ACCELERATION
## Miles per Hour-Second
## to
## Kilometers per Hour-Second

| mi/(h·s) | km/(h·s) | mi/(h·s) | km/(h·s) |
|---|---|---|---|
| 1 | 1.609300 | 51 | 82.074300 |
| 2 | 3.218600 | 52 | 83.683600 |
| 3 | 4.827900 | 53 | 85.292900 |
| 4 | 6.437200 | 54 | 86.902200 |
| 5 | 8.046500 | 55 | 88.511500 |
| 6 | 9.655800 | 56 | 90.120800 |
| 7 | 11.265100 | 57 | 91.730100 |
| 8 | 12.874400 | 58 | 93.339400 |
| 9 | 14.483700 | 59 | 94.948700 |
| 10 | 16.093000 | 60 | 96.558000 |
| 11 | 17.702300 | 61 | 98.167300 |
| 12 | 19.311600 | 62 | 99.776600 |
| 13 | 20.920900 | 63 | 101.385900 |
| 14 | 22.530200 | 64 | 102.995200 |
| 15 | 24.139500 | 65 | 104.604500 |
| 16 | 25.748800 | 66 | 106.213800 |
| 17 | 27.358100 | 67 | 107.823100 |
| 18 | 28.967400 | 68 | 109.432400 |
| 19 | 30.576700 | 69 | 111.041700 |
| 20 | 32.186000 | 70 | 112.651000 |
| 21 | 33.795300 | 71 | 114.260300 |
| 22 | 35.404600 | 72 | 115.869600 |
| 23 | 37.013900 | 73 | 117.478900 |
| 24 | 38.623200 | 74 | 119.088200 |
| 25 | 40.232500 | 75 | 120.697500 |
| 26 | 41.841800 | 76 | 122.306800 |
| 27 | 43.451100 | 77 | 123.916100 |
| 28 | 45.060400 | 78 | 125.525400 |
| 29 | 46.669700 | 79 | 127.134700 |
| 30 | 48.279000 | 80 | 128.744000 |
| 31 | 49.888300 | 81 | 130.353300 |
| 32 | 51.497600 | 82 | 131.962600 |
| 33 | 53.106900 | 83 | 133.571900 |
| 34 | 54.716200 | 84 | 135.181200 |
| 35 | 56.325500 | 85 | 136.790500 |
| 36 | 57.934800 | 86 | 138.399800 |
| 37 | 59.544100 | 87 | 140.009100 |
| 38 | 61.153400 | 88 | 141.618400 |
| 39 | 62.762700 | 89 | 143.227700 |
| 40 | 64.372000 | 90 | 144.837000 |
| 41 | 65.981300 | 91 | 146.446300 |
| 42 | 67.590600 | 92 | 148.055600 |
| 43 | 69.199900 | 93 | 149.664900 |
| 44 | 70.809200 | 94 | 151.274200 |
| 45 | 72.418500 | 95 | 152.883500 |
| 46 | 74.027800 | 96 | 154.492800 |
| 47 | 75.637100 | 97 | 156.102100 |
| 48 | 77.246400 | 98 | 157.711400 |
| 49 | 78.855700 | 99 | 159.320700 |
| 50 | 80.465000 | 100 | 160.930000 |

## ACCELERATION
### Kilometers per Hour-Second
### to
### Miles per Hour-Second

| km/(h·s) | mi/(h·s) | km/(h·s) | mi/(h·s) |
|---|---|---|---|
| 1 | .621400 | 51 | 31.691400 |
| 2 | 1.242800 | 52 | 32.312800 |
| 3 | 1.864200 | 53 | 32.934200 |
| 4 | 2.485600 | 54 | 33.555600 |
| 5 | 3.107000 | 55 | 34.177000 |
| 6 | 3.728400 | 56 | 34.798400 |
| 7 | 4.349800 | 57 | 35.419800 |
| 8 | 4.971200 | 58 | 36.041200 |
| 9 | 5.592600 | 59 | 36.662600 |
| 10 | 6.214000 | 60 | 37.284000 |
| 11 | 6.835400 | 61 | 37.905400 |
| 12 | 7.456800 | 62 | 38.526800 |
| 13 | 8.078200 | 63 | 39.148200 |
| 14 | 8.699600 | 64 | 39.769600 |
| 15 | 9.321000 | 65 | 40.391000 |
| 16 | 9.942400 | 66 | 41.012400 |
| 17 | 10.563800 | 67 | 41.633800 |
| 18 | 11.185200 | 68 | 42.255200 |
| 19 | 11.806600 | 69 | 42.876600 |
| 20 | 12.428000 | 70 | 43.498000 |
| 21 | 13.049400 | 71 | 44.119400 |
| 22 | 13.670800 | 72 | 44.740800 |
| 23 | 14.292200 | 73 | 45.362200 |
| 24 | 14.913600 | 74 | 45.983600 |
| 25 | 15.535000 | 75 | 46.605000 |
| 26 | 16.156400 | 76 | 47.226400 |
| 27 | 16.777800 | 77 | 47.847800 |
| 28 | 17.399200 | 78 | 48.469200 |
| 29 | 18.020600 | 79 | 49.090600 |
| 30 | 18.642000 | 80 | 49.712000 |
| 31 | 19.263400 | 81 | 50.333400 |
| 32 | 19.884800 | 82 | 50.954800 |
| 33 | 20.506200 | 83 | 51.576200 |
| 34 | 21.127600 | 84 | 52.197600 |
| 35 | 21.749000 | 85 | 52.819000 |
| 36 | 22.370400 | 86 | 53.440400 |
| 37 | 22.991800 | 87 | 54.061800 |
| 38 | 23.613200 | 88 | 54.683200 |
| 39 | 24.234600 | 89 | 55.304600 |
| 40 | 24.856000 | 90 | 55.926000 |
| 41 | 25.477400 | 91 | 56.547400 |
| 42 | 26.098800 | 92 | 57.168800 |
| 43 | 26.720200 | 93 | 57.790200 |
| 44 | 27.341600 | 94 | 58.411600 |
| 45 | 27.963000 | 95 | 59.033000 |
| 46 | 28.584400 | 96 | 59.654400 |
| 47 | 29.205800 | 97 | 60.275800 |
| 48 | 29.827200 | 98 | 60.897200 |
| 49 | 30.448600 | 99 | 61.518600 |
| 50 | 31.070000 | 100 | 62.140000 |

## PRESSURE
## Pounds per Square Inch
## to
## Kilograms per Square Centimeter

| lb/in² | kg/cm² | lb/in² | kg/cm² |
|---|---|---|---|
| 1 | .070307 | 51 | 3.585657 |
| 2 | .140614 | 52 | 3.655964 |
| 3 | .210921 | 53 | 3.726271 |
| 4 | .281228 | 54 | 3.796578 |
| 5 | .351535 | 55 | 3.866885 |
| 6 | .421842 | 56 | 3.937192 |
| 7 | .492149 | 57 | 4.007499 |
| 8 | .562456 | 58 | 4.077806 |
| 9 | .632763 | 59 | 4.148113 |
| 10 | .703070 | 60 | 4.218420 |
| 11 | .773377 | 61 | 4.288727 |
| 12 | .843684 | 62 | 4.359034 |
| 13 | .913991 | 63 | 4.429341 |
| 14 | .984298 | 64 | 4.499648 |
| 15 | 1.054605 | 65 | 4.569955 |
| 16 | 1.124912 | 66 | 4.640262 |
| 17 | 1.195219 | 67 | 4.710569 |
| 18 | 1.265526 | 68 | 4.780876 |
| 19 | 1.335833 | 69 | 4.851183 |
| 20 | 1.406140 | 70 | 4.921490 |
| 21 | 1.476447 | 71 | 4.991797 |
| 22 | 1.546754 | 72 | 5.062104 |
| 23 | 1.617061 | 73 | 5.132411 |
| 24 | 1.687368 | 74 | 5.202718 |
| 25 | 1.757675 | 75 | 5.273025 |
| 26 | 1.827982 | 76 | 5.343332 |
| 27 | 1.898289 | 77 | 5.413639 |
| 28 | 1.968596 | 78 | 5.483946 |
| 29 | 2.038903 | 79 | 5.554253 |
| 30 | 2.109210 | 80 | 5.624560 |
| 31 | 2.179517 | 81 | 5.694867 |
| 32 | 2.249824 | 82 | 5.765174 |
| 33 | 2.320131 | 83 | 5.835481 |
| 34 | 2.390438 | 84 | 5.905788 |
| 35 | 2.460745 | 85 | 5.976095 |
| 36 | 2.531052 | 86 | 6.046402 |
| 37 | 2.601359 | 87 | 6.116709 |
| 38 | 2.671666 | 88 | 6.187016 |
| 39 | 2.741973 | 89 | 6.257323 |
| 40 | 2.812280 | 90 | 6.327630 |
| 41 | 2.882587 | 91 | 6.397937 |
| 42 | 2.952894 | 92 | 6.468244 |
| 43 | 3.023201 | 93 | 6.538551 |
| 44 | 3.093508 | 94 | 6.608858 |
| 45 | 3.163815 | 95 | 6.679165 |
| 46 | 3.234122 | 96 | 6.749472 |
| 47 | 3.304429 | 97 | 6.819779 |
| 48 | 3.374736 | 98 | 6.890086 |
| 49 | 3.445043 | 99 | 6.960393 |
| 50 | 3.515350 | 100 | 7.030700 |

# PRESSURE
## Kilograms per Square Centimeter to Pounds per Square Inch

| kg/cm² | lb/in² | kg/cm² | lb/in² |
|---|---|---|---|
| 1 | 14.223340 | 51 | 725.390340 |
| 2 | 28.446680 | 52 | 739.613680 |
| 3 | 42.670020 | 53 | 753.837020 |
| 4 | 56.893360 | 54 | 768.060360 |
| 5 | 71.116700 | 55 | 782.283700 |
| 6 | 85.340040 | 56 | 796.507040 |
| 7 | 99.563380 | 57 | 810.730380 |
| 8 | 113.786720 | 58 | 824.953720 |
| 9 | 128.010060 | 59 | 839.177060 |
| 10 | 142.233400 | 60 | 853.400400 |
| 11 | 156.456740 | 61 | 867.623740 |
| 12 | 170.680080 | 62 | 881.847080 |
| 13 | 184.903420 | 63 | 896.070420 |
| 14 | 199.126760 | 64 | 910.293760 |
| 15 | 213.350100 | 65 | 924.517100 |
| 16 | 227.573440 | 66 | 938.740440 |
| 17 | 241.796780 | 67 | 952.963780 |
| 18 | 256.020120 | 68 | 967.187120 |
| 19 | 270.243460 | 69 | 981.410460 |
| 20 | 284.466800 | 70 | 995.633800 |
| 21 | 298.690140 | 71 | 1009.857140 |
| 22 | 312.913480 | 72 | 1024.080480 |
| 23 | 327.136820 | 73 | 1038.303820 |
| 24 | 341.360160 | 74 | 1052.527160 |
| 25 | 355.583500 | 75 | 1066.750500 |
| 26 | 369.806840 | 76 | 1080.973840 |
| 27 | 384.030180 | 77 | 1095.197180 |
| 28 | 398.253520 | 78 | 1109.420520 |
| 29 | 412.476860 | 79 | 1123.643860 |
| 30 | 426.700200 | 80 | 1137.867200 |
| 31 | 440.923540 | 81 | 1152.090540 |
| 32 | 455.146880 | 82 | 1166.313880 |
| 33 | 469.370220 | 83 | 1180.537220 |
| 34 | 483.593560 | 84 | 1194.760560 |
| 35 | 497.816900 | 85 | 1208.983900 |
| 36 | 512.040240 | 86 | 1223.207240 |
| 37 | 526.263580 | 87 | 1237.430580 |
| 38 | 540.486920 | 88 | 1251.653920 |
| 39 | 554.710260 | 89 | 1265.877260 |
| 40 | 568.933600 | 90 | 1280.100600 |
| 41 | 583.156940 | 91 | 1294.323940 |
| 42 | 597.380280 | 92 | 1308.547280 |
| 43 | 611.603620 | 93 | 1322.770620 |
| 44 | 625.826960 | 94 | 1336.993960 |
| 45 | 640.050300 | 95 | 1351.217300 |
| 46 | 654.273640 | 96 | 1365.440640 |
| 47 | 668.496980 | 97 | 1379.663980 |
| 48 | 682.720320 | 98 | 1393.887320 |
| 49 | 696.943660 | 99 | 1408.110660 |
| 50 | 711.167000 | 100 | 1422.334000 |

## PRESSURE
## Pounds per Square Inch to Kilopascals

| lb/in² | kPa | lb/in² | kPa |
|---|---|---|---|
| 1 | 6.894757 | 51 | 351.632607 |
| 2 | 13.789514 | 52 | 358.527364 |
| 3 | 20.684271 | 53 | 365.422121 |
| 4 | 27.579028 | 54 | 372.316878 |
| 5 | 34.473785 | 55 | 379.211635 |
| 6 | 41.368542 | 56 | 386.106392 |
| 7 | 48.263299 | 57 | 393.001149 |
| 8 | 55.158056 | 58 | 399.895906 |
| 9 | 62.052813 | 59 | 406.790663 |
| 10 | 68.947570 | 60 | 413.685420 |
| 11 | 75.842327 | 61 | 420.580177 |
| 12 | 82.737084 | 62 | 427.474934 |
| 13 | 89.631841 | 63 | 434.369691 |
| 14 | 96.526598 | 64 | 441.264448 |
| 15 | 103.421355 | 65 | 448.159205 |
| 16 | 110.316112 | 66 | 455.053962 |
| 17 | 117.210869 | 67 | 461.948719 |
| 18 | 124.105626 | 68 | 468.843476 |
| 19 | 131.000383 | 69 | 475.738233 |
| 20 | 137.895140 | 70 | 482.632990 |
| 21 | 144.789897 | 71 | 489.527747 |
| 22 | 151.684654 | 72 | 496.422504 |
| 23 | 158.579411 | 73 | 503.317261 |
| 24 | 165.474168 | 74 | 510.212018 |
| 25 | 172.368925 | 75 | 517.106775 |
| 26 | 179.263682 | 76 | 524.001532 |
| 27 | 186.158439 | 77 | 530.896289 |
| 28 | 193.053196 | 78 | 537.791046 |
| 29 | 199.947953 | 79 | 544.685803 |
| 30 | 206.842710 | 80 | 551.580560 |
| 31 | 213.737467 | 81 | 558.475317 |
| 32 | 220.632224 | 82 | 565.370074 |
| 33 | 227.526981 | 83 | 572.264831 |
| 34 | 234.421738 | 84 | 579.159588 |
| 35 | 241.316495 | 85 | 586.054345 |
| 36 | 248.211252 | 86 | 592.949102 |
| 37 | 255.106009 | 87 | 599.843859 |
| 38 | 262.000766 | 88 | 606.738616 |
| 39 | 268.895523 | 89 | 613.633373 |
| 40 | 275.790280 | 90 | 620.528130 |
| 41 | 282.685037 | 91 | 627.422887 |
| 42 | 289.579794 | 92 | 634.317644 |
| 43 | 296.474551 | 93 | 641.212401 |
| 44 | 303.369308 | 94 | 648.107158 |
| 45 | 310.264065 | 95 | 655.001915 |
| 46 | 317.158822 | 96 | 661.896672 |
| 47 | 324.053579 | 97 | 668.791429 |
| 48 | 330.948336 | 98 | 675.686186 |
| 49 | 337.843093 | 99 | 682.580943 |
| 50 | 344.737850 | 100 | 689.475700 |

## PRESSURE
## Kilopascals
## to
## Pounds per Square Inch

| kPa | lb/in² | kPa | lb/in² |
|---|---|---|---|
| 1 | .145038 | 51 | 7.396938 |
| 2 | .290076 | 52 | 7.541976 |
| 3 | .435114 | 53 | 7.687014 |
| 4 | .580152 | 54 | 7.832052 |
| 5 | .725190 | 55 | 7.977090 |
| 6 | .870228 | 56 | 8.122128 |
| 7 | 1.015266 | 57 | 8.267166 |
| 8 | 1.160304 | 58 | 8.412204 |
| 9 | 1.305342 | 59 | 8.557242 |
| 10 | 1.450380 | 60 | 8.702280 |
| 11 | 1.595418 | 61 | 8.847318 |
| 12 | 1.740456 | 62 | 8.992356 |
| 13 | 1.885494 | 63 | 9.137394 |
| 14 | 2.030532 | 64 | 9.282432 |
| 15 | 2.175570 | 65 | 9.427470 |
| 16 | 2.320608 | 66 | 9.572508 |
| 17 | 2.465646 | 67 | 9.717546 |
| 18 | 2.610684 | 68 | 9.862584 |
| 19 | 2.755722 | 69 | 10.007622 |
| 20 | 2.900760 | 70 | 10.152660 |
| 21 | 3.045798 | 71 | 10.297698 |
| 22 | 3.190836 | 72 | 10.442736 |
| 23 | 3.335874 | 73 | 10.587774 |
| 24 | 3.480912 | 74 | 10.732812 |
| 25 | 3.625950 | 75 | 10.877850 |
| 26 | 3.770988 | 76 | 11.022888 |
| 27 | 3.916026 | 77 | 11.167926 |
| 28 | 4.061064 | 78 | 11.312964 |
| 29 | 4.206102 | 79 | 11.458002 |
| 30 | 4.351140 | 80 | 11.603040 |
| 31 | 4.496178 | 81 | 11.748078 |
| 32 | 4.641216 | 82 | 11.893116 |
| 33 | 4.786254 | 83 | 12.038154 |
| 34 | 4.931292 | 84 | 12.183192 |
| 35 | 5.076330 | 85 | 12.328230 |
| 36 | 5.221368 | 86 | 12.473268 |
| 37 | 5.366406 | 87 | 12.618306 |
| 38 | 5.511444 | 88 | 12.763344 |
| 39 | 5.656482 | 89 | 12.908382 |
| 40 | 5.801520 | 90 | 13.053420 |
| 41 | 5.946558 | 91 | 13.198458 |
| 42 | 6.091596 | 92 | 13.343496 |
| 43 | 6.236634 | 93 | 13.488534 |
| 44 | 6.381672 | 94 | 13.633572 |
| 45 | 6.526710 | 95 | 13.778610 |
| 46 | 6.671748 | 96 | 13.923648 |
| 47 | 6.816786 | 97 | 14.068686 |
| 48 | 6.961824 | 98 | 14.213724 |
| 49 | 7.106862 | 99 | 14.358762 |
| 50 | 7.251900 | 100 | 14.503800 |

## PRESSURE
### Pounds per Square Foot
### to
### Kilograms per Square Meter

| lb/ft² | kg/m² | lb/ft² | kg/m² |
|---|---|---|---|
| 1 | 4.882423 | 51 | 249.003573 |
| 2 | 9.764846 | 52 | 253.885996 |
| 3 | 14.647269 | 53 | 258.768419 |
| 4 | 19.529692 | 54 | 263.650842 |
| 5 | 24.412115 | 55 | 268.533265 |
| 6 | 29.294538 | 56 | 273.415688 |
| 7 | 34.176961 | 57 | 278.298111 |
| 8 | 39.059384 | 58 | 283.180534 |
| 9 | 43.941807 | 59 | 288.062957 |
| 10 | 48.824230 | 60 | 292.945380 |
| 11 | 53.706653 | 61 | 297.827803 |
| 12 | 58.589076 | 62 | 302.710226 |
| 13 | 63.471499 | 63 | 307.592649 |
| 14 | 68.353922 | 64 | 312.475072 |
| 15 | 73.236345 | 65 | 317.357495 |
| 16 | 78.118768 | 66 | 322.239918 |
| 17 | 83.001191 | 67 | 327.122341 |
| 18 | 87.883614 | 68 | 332.004764 |
| 19 | 92.766037 | 69 | 336.887187 |
| 20 | 97.648460 | 70 | 341.769610 |
| 21 | 102.530883 | 71 | 346.652033 |
| 22 | 107.413306 | 72 | 351.534456 |
| 23 | 112.295729 | 73 | 356.416879 |
| 24 | 117.178152 | 74 | 361.299302 |
| 25 | 122.060575 | 75 | 366.181725 |
| 26 | 126.942998 | 76 | 371.064148 |
| 27 | 131.825421 | 77 | 375.946571 |
| 28 | 136.707844 | 78 | 380.828994 |
| 29 | 141.590267 | 79 | 385.711417 |
| 30 | 146.472690 | 80 | 390.593840 |
| 31 | 151.355113 | 81 | 395.476263 |
| 32 | 156.237536 | 82 | 400.358686 |
| 33 | 161.119959 | 83 | 405.241109 |
| 34 | 166.002382 | 84 | 410.123532 |
| 35 | 170.884805 | 85 | 415.005955 |
| 36 | 175.767228 | 86 | 419.888378 |
| 37 | 180.649651 | 87 | 424.770801 |
| 38 | 185.532074 | 88 | 429.653224 |
| 39 | 190.414497 | 89 | 434.535647 |
| 40 | 195.296920 | 90 | 439.418070 |
| 41 | 200.179343 | 91 | 444.300493 |
| 42 | 205.061766 | 92 | 449.182916 |
| 43 | 209.944189 | 93 | 454.065339 |
| 44 | 214.826612 | 94 | 458.947762 |
| 45 | 219.709035 | 95 | 463.830185 |
| 46 | 224.591458 | 96 | 468.712608 |
| 47 | 229.473881 | 97 | 473.595031 |
| 48 | 234.356304 | 98 | 478.477454 |
| 49 | 239.238727 | 99 | 483.359877 |
| 50 | 244.121150 | 100 | 488.242300 |

## PRESSURE
### Kilograms per Square Meter
### to
### Pounds per Square Foot

| kg/m² | lb/ft² | kg/m² | lb/ft² |
|---|---|---|---|
| 1 | .204816 | 51 | 10.445616 |
| 2 | .409632 | 52 | 10.650432 |
| 3 | .614448 | 53 | 10.855248 |
| 4 | .819264 | 54 | 11.060064 |
| 5 | 1.024080 | 55 | 11.264880 |
| 6 | 1.228896 | 56 | 11.469696 |
| 7 | 1.433712 | 57 | 11.674512 |
| 8 | 1.638528 | 58 | 11.879328 |
| 9 | 1.843344 | 59 | 12.084144 |
| 10 | 2.048160 | 60 | 12.288960 |
| 11 | 2.252976 | 61 | 12.493776 |
| 12 | 2.457792 | 62 | 12.698592 |
| 13 | 2.662608 | 63 | 12.903408 |
| 14 | 2.867424 | 64 | 13.108224 |
| 15 | 3.072240 | 65 | 13.313040 |
| 16 | 3.277056 | 66 | 13.517856 |
| 17 | 3.481872 | 67 | 13.722672 |
| 18 | 3.686688 | 68 | 13.927488 |
| 19 | 3.891504 | 69 | 14.132304 |
| 20 | 4.096320 | 70 | 14.337120 |
| 21 | 4.301136 | 71 | 14.541936 |
| 22 | 4.505952 | 72 | 14.746752 |
| 23 | 4.710768 | 73 | 14.951568 |
| 24 | 4.915584 | 74 | 15.156384 |
| 25 | 5.120400 | 75 | 15.361200 |
| 26 | 5.325216 | 76 | 15.566016 |
| 27 | 5.530032 | 77 | 15.770832 |
| 28 | 5.734848 | 78 | 15.975648 |
| 29 | 5.939664 | 79 | 16.180464 |
| 30 | 6.144480 | 80 | 16.385280 |
| 31 | 6.349296 | 81 | 16.590096 |
| 32 | 6.554112 | 82 | 16.794912 |
| 33 | 6.758928 | 83 | 16.999728 |
| 34 | 6.963744 | 84 | 17.204544 |
| 35 | 7.168560 | 85 | 17.409360 |
| 36 | 7.373376 | 86 | 17.614176 |
| 37 | 7.578192 | 87 | 17.818992 |
| 38 | 7.783008 | 88 | 18.023808 |
| 39 | 7.987824 | 89 | 18.228624 |
| 40 | 8.192640 | 90 | 18.433440 |
| 41 | 8.397456 | 91 | 18.638256 |
| 42 | 8.602272 | 92 | 18.843072 |
| 43 | 8.807088 | 93 | 19.047888 |
| 44 | 9.011904 | 94 | 19.252704 |
| 45 | 9.216720 | 95 | 19.457520 |
| 46 | 9.421536 | 96 | 19.662336 |
| 47 | 9.626352 | 97 | 19.867152 |
| 48 | 9.831168 | 98 | 20.071968 |
| 49 | 10.035984 | 99 | 20.276784 |
| 50 | 10.240800 | 100 | 20.481600 |

## PRESSURE
## Kilograms per Square Meter
## to
## Pascals

| kg/m² | Pa | kg/m² | Pa |
|---|---|---|---|
| 1 | 9.806650 | 51 | 500.139150 |
| 2 | 19.613300 | 52 | 509.945800 |
| 3 | 29.419950 | 53 | 519.752450 |
| 4 | 39.226600 | 54 | 529.559100 |
| 5 | 49.033250 | 55 | 539.365750 |
| 6 | 58.839900 | 56 | 549.172400 |
| 7 | 68.646550 | 57 | 558.979050 |
| 8 | 78.453200 | 58 | 568.785700 |
| 9 | 88.259850 | 59 | 578.592350 |
| 10 | 98.066500 | 60 | 588.399000 |
| 11 | 107.873150 | 61 | 598.205650 |
| 12 | 117.679800 | 62 | 608.012300 |
| 13 | 127.486450 | 63 | 617.818950 |
| 14 | 137.293100 | 64 | 627.625600 |
| 15 | 147.099750 | 65 | 637.432250 |
| 16 | 156.906400 | 66 | 647.238900 |
| 17 | 166.713050 | 67 | 657.045550 |
| 18 | 176.519700 | 68 | 666.852200 |
| 19 | 186.326350 | 69 | 676.658850 |
| 20 | 196.133000 | 70 | 686.465500 |
| 21 | 205.939650 | 71 | 696.272150 |
| 22 | 215.746300 | 72 | 706.078800 |
| 23 | 225.552950 | 73 | 715.885450 |
| 24 | 235.359600 | 74 | 725.692100 |
| 25 | 245.166250 | 75 | 735.498750 |
| 26 | 254.972900 | 76 | 745.305400 |
| 27 | 264.779550 | 77 | 755.112050 |
| 28 | 274.586200 | 78 | 764.918700 |
| 29 | 284.392850 | 79 | 774.725350 |
| 30 | 294.199500 | 80 | 784.532000 |
| 31 | 304.006150 | 81 | 794.338650 |
| 32 | 313.812800 | 82 | 804.145300 |
| 33 | 323.619450 | 83 | 813.951950 |
| 34 | 333.426100 | 84 | 823.758600 |
| 35 | 343.232750 | 85 | 833.565250 |
| 36 | 353.039400 | 86 | 843.371900 |
| 37 | 362.846050 | 87 | 853.178550 |
| 38 | 372.652700 | 88 | 862.985200 |
| 39 | 382.459350 | 89 | 872.791850 |
| 40 | 392.266000 | 90 | 882.598500 |
| 41 | 402.072650 | 91 | 892.405150 |
| 42 | 411.879300 | 92 | 902.211800 |
| 43 | 421.685950 | 93 | 912.018450 |
| 44 | 431.492600 | 94 | 921.825100 |
| 45 | 441.299250 | 95 | 931.631750 |
| 46 | 451.105900 | 96 | 941.438400 |
| 47 | 460.912550 | 97 | 951.245050 |
| 48 | 470.719200 | 98 | 961.051700 |
| 49 | 480.525850 | 99 | 970.858350 |
| 50 | 490.332500 | 100 | 980.665000 |

## PRESSURE
## Pascals
## to
## Kilograms per Square Meter

| Pa | kg/m² | Pa | kg/m² |
|---|---|---|---|
| 1 | .101972 | 51 | 5.200572 |
| 2 | .203944 | 52 | 5.302544 |
| 3 | .305916 | 53 | 5.404516 |
| 4 | .407888 | 54 | 5.506488 |
| 5 | .509860 | 55 | 5.608460 |
| 6 | .611832 | 56 | 5.710432 |
| 7 | .713804 | 57 | 5.812404 |
| 8 | .815776 | 58 | 5.914376 |
| 9 | .917748 | 59 | 6.016348 |
| 10 | 1.019720 | 60 | 6.118320 |
| 11 | 1.121692 | 61 | 6.220292 |
| 12 | 1.223664 | 62 | 6.322264 |
| 13 | 1.325636 | 63 | 6.424236 |
| 14 | 1.427608 | 64 | 6.526208 |
| 15 | 1.529580 | 65 | 6.628180 |
| 16 | 1.631552 | 66 | 6.730152 |
| 17 | 1.733524 | 67 | 6.832124 |
| 18 | 1.835496 | 68 | 6.934096 |
| 19 | 1.937468 | 69 | 7.036068 |
| 20 | 2.039440 | 70 | 7.138040 |
| 21 | 2.141412 | 71 | 7.240012 |
| 22 | 2.243384 | 72 | 7.341984 |
| 23 | 2.345356 | 73 | 7.443956 |
| 24 | 2.447328 | 74 | 7.545928 |
| 25 | 2.549300 | 75 | 7.647900 |
| 26 | 2.651272 | 76 | 7.749872 |
| 27 | 2.753244 | 77 | 7.851844 |
| 28 | 2.855216 | 78 | 7.953816 |
| 29 | 2.957188 | 79 | 8.055788 |
| 30 | 3.059160 | 80 | 8.157760 |
| 31 | 3.161132 | 81 | 8.259732 |
| 32 | 3.263104 | 82 | 8.361704 |
| 33 | 3.365076 | 83 | 8.463676 |
| 34 | 3.467048 | 84 | 8.565648 |
| 35 | 3.569020 | 85 | 8.667620 |
| 36 | 3.670992 | 86 | 8.769592 |
| 37 | 3.772964 | 87 | 8.871564 |
| 38 | 3.874936 | 88 | 8.973536 |
| 39 | 3.976908 | 89 | 9.075508 |
| 40 | 4.078880 | 90 | 9.177480 |
| 41 | 4.180852 | 91 | 9.279452 |
| 42 | 4.282824 | 92 | 9.381424 |
| 43 | 4.384796 | 93 | 9.483396 |
| 44 | 4.486768 | 94 | 9.585368 |
| 45 | 4.588740 | 95 | 9.687340 |
| 46 | 4.690712 | 96 | 9.789312 |
| 47 | 4.792684 | 97 | 9.891284 |
| 48 | 4.894656 | 98 | 9.993256 |
| 49 | 4.996628 | 99 | 10.095228 |
| 50 | 5.098600 | 100 | 10.197200 |

## PRESSURE
## Pounds per Square Foot
## to
## Pascals

| lb/ft² | Pa | lb/ft² | Pa |
|---|---|---|---|
| 1 | 47.880257 | 51 | 2441.893107 |
| 2 | 95.760514 | 52 | 2489.773364 |
| 3 | 143.640771 | 53 | 2537.653621 |
| 4 | 191.521028 | 54 | 2585.533878 |
| 5 | 239.401285 | 55 | 2633.414135 |
| 6 | 287.281542 | 56 | 2681.294392 |
| 7 | 335.161799 | 57 | 2729.174649 |
| 8 | 383.042056 | 58 | 2777.054906 |
| 9 | 430.922313 | 59 | 2824.935163 |
| 10 | 478.802570 | 60 | 2872.815420 |
| 11 | 526.682827 | 61 | 2920.695677 |
| 12 | 574.563084 | 62 | 2968.575934 |
| 13 | 622.443341 | 63 | 3016.456191 |
| 14 | 670.323598 | 64 | 3064.336448 |
| 15 | 718.203855 | 65 | 3112.216705 |
| 16 | 766.084112 | 66 | 3160.096962 |
| 17 | 813.964369 | 67 | 3207.977219 |
| 18 | 861.844626 | 68 | 3255.857476 |
| 19 | 909.724883 | 69 | 3303.737733 |
| 20 | 957.605140 | 70 | 3351.617990 |
| 21 | 1005.485397 | 71 | 3399.498247 |
| 22 | 1053.365654 | 72 | 3447.378504 |
| 23 | 1101.245911 | 73 | 3495.258761 |
| 24 | 1149.126168 | 74 | 3543.139018 |
| 25 | 1197.006425 | 75 | 3591.019275 |
| 26 | 1244.886682 | 76 | 3638.899532 |
| 27 | 1292.766939 | 77 | 3686.779789 |
| 28 | 1340.647196 | 78 | 3734.660046 |
| 29 | 1388.527453 | 79 | 3782.540303 |
| 30 | 1436.407710 | 80 | 3830.420560 |
| 31 | 1484.287967 | 81 | 3878.300817 |
| 32 | 1532.168224 | 82 | 3926.181074 |
| 33 | 1580.048481 | 83 | 3974.061331 |
| 34 | 1627.928738 | 84 | 4021.941588 |
| 35 | 1675.808995 | 85 | 4069.821845 |
| 36 | 1723.689252 | 86 | 4117.702102 |
| 37 | 1771.569509 | 87 | 4165.582359 |
| 38 | 1819.449766 | 88 | 4213.462616 |
| 39 | 1867.330023 | 89 | 4261.342873 |
| 40 | 1915.210280 | 90 | 4309.223130 |
| 41 | 1963.090537 | 91 | 4357.103387 |
| 42 | 2010.970794 | 92 | 4404.983644 |
| 43 | 2058.851051 | 93 | 4452.863901 |
| 44 | 2106.731308 | 94 | 4500.744158 |
| 45 | 2154.611565 | 95 | 4548.624415 |
| 46 | 2202.491822 | 96 | 4596.504672 |
| 47 | 2250.372079 | 97 | 4644.384929 |
| 48 | 2298.252336 | 98 | 4692.265186 |
| 49 | 2346.132593 | 99 | 4740.145443 |
| 50 | 2394.012850 | 100 | 4788.025700 |

# PRESSURE
## Pascals
## to
## Pounds per Square Foot

| Pa | lb/ft² | Pa | lb/ft² |
|---|---|---|---|
| 1 | .020885 | 51 | 1.065135 |
| 2 | .041770 | 52 | 1.086020 |
| 3 | .062655 | 53 | 1.106905 |
| 4 | .083540 | 54 | 1.127790 |
| 5 | .104425 | 55 | 1.148675 |
| 6 | .125310 | 56 | 1.169560 |
| 7 | .146195 | 57 | 1.190445 |
| 8 | .167080 | 58 | 1.211330 |
| 9 | .187965 | 59 | 1.232215 |
| 10 | .208850 | 60 | 1.253100 |
| 11 | .229735 | 61 | 1.273985 |
| 12 | .250620 | 62 | 1.294870 |
| 13 | .271505 | 63 | 1.315755 |
| 14 | .292390 | 64 | 1.336640 |
| 15 | .313275 | 65 | 1.357525 |
| 16 | .334160 | 66 | 1.378410 |
| 17 | .355045 | 67 | 1.399295 |
| 18 | .375930 | 68 | 1.420180 |
| 19 | .396815 | 69 | 1.441065 |
| 20 | .417700 | 70 | 1.461950 |
| 21 | .438585 | 71 | 1.482835 |
| 22 | .459470 | 72 | 1.503720 |
| 23 | .480355 | 73 | 1.524605 |
| 24 | .501240 | 74 | 1.545490 |
| 25 | .522125 | 75 | 1.566375 |
| 26 | .543010 | 76 | 1.587260 |
| 27 | .563895 | 77 | 1.608145 |
| 28 | .584780 | 78 | 1.629030 |
| 29 | .605665 | 79 | 1.649915 |
| 30 | .626550 | 80 | 1.670800 |
| 31 | .647435 | 81 | 1.691685 |
| 32 | .668320 | 82 | 1.712570 |
| 33 | .689205 | 83 | 1.733455 |
| 34 | .710090 | 84 | 1.754340 |
| 35 | .730975 | 85 | 1.775225 |
| 36 | .751860 | 86 | 1.796110 |
| 37 | .772745 | 87 | 1.816995 |
| 38 | .793630 | 88 | 1.837880 |
| 39 | .814515 | 89 | 1.858765 |
| 40 | .835400 | 90 | 1.879650 |
| 41 | .856285 | 91 | 1.900535 |
| 42 | .877170 | 92 | 1.921420 |
| 43 | .898055 | 93 | 1.942305 |
| 44 | .918940 | 94 | 1.963190 |
| 45 | .939825 | 95 | 1.984075 |
| 46 | .960710 | 96 | 2.004960 |
| 47 | .981595 | 97 | 2.025845 |
| 48 | 1.002480 | 98 | 2.046730 |
| 49 | 1.023365 | 99 | 2.067615 |
| 50 | 1.044250 | 100 | 2.088500 |

## PRESSURE
### Inches of Mercury to Kilopascals

| in Hg | kPa | in Hg | kPa |
|---|---|---|---|
| 25.0 | 84.659725 | 30.0 | 101.591670 |
| 25.1 | 84.998364 | 30.1 | 101.930309 |
| 25.2 | 85.337003 | 30.2 | 102.268948 |
| 25.3 | 85.675642 | 30.3 | 102.607587 |
| 25.4 | 86.014281 | 30.4 | 102.946226 |
| 25.5 | 86.352920 | 30.5 | 103.284865 |
| 25.6 | 86.691558 | 30.6 | 103.623503 |
| 25.7 | 87.030197 | 30.7 | 103.962142 |
| 25.8 | 87.368836 | 30.8 | 104.300781 |
| 25.9 | 87.707475 | 30.9 | 104.639420 |
| 26.0 | 88.046114 | 31.0 | 104.978059 |
| 26.1 | 88.384753 | 31.1 | 105.316698 |
| 26.2 | 88.723392 | 31.2 | 105.655337 |
| 26.3 | 89.062031 | 31.3 | 105.993976 |
| 26.4 | 89.400670 | 31.4 | 106.332615 |
| 26.5 | 89.739309 | 31.5 | 106.671254 |
| 26.6 | 90.077947 | 31.6 | 107.009892 |
| 26.7 | 90.416586 | 31.7 | 107.348531 |
| 26.8 | 90.755225 | 31.8 | 107.687170 |
| 26.9 | 91.093864 | 31.9 | 108.025809 |
| 27.0 | 91.432503 | 32.0 | 108.364448 |
| 27.1 | 91.771142 | 32.1 | 108.703087 |
| 27.2 | 92.109781 | 32.2 | 109.041726 |
| 27.3 | 92.448420 | 32.3 | 109.380365 |
| 27.4 | 92.787059 | 32.4 | 109.719004 |
| 27.5 | 93.125698 | 32.5 | 110.057643 |
| 27.6 | 93.464336 | 32.6 | 110.396281 |
| 27.7 | 93.802975 | 32.7 | 110.734920 |
| 27.8 | 94.141614 | 32.8 | 111.073559 |
| 27.9 | 94.480253 | 32.9 | 111.412198 |
| 28.0 | 94.818892 | 33.0 | 111.750837 |
| 28.1 | 95.157531 | 33.1 | 112.089476 |
| 28.2 | 95.496170 | 33.2 | 112.428115 |
| 28.3 | 95.834809 | 33.3 | 112.766754 |
| 28.4 | 96.173448 | 33.4 | 113.105393 |
| 28.5 | 96.512087 | 33.5 | 113.444032 |
| 28.6 | 96.850725 | 33.6 | 113.782670 |
| 28.7 | 97.189364 | 33.7 | 114.121309 |
| 28.8 | 97.528003 | 33.8 | 114.459948 |
| 28.9 | 97.866642 | 33.9 | 114.798587 |
| 29.0 | 98.205281 | 34.0 | 115.137226 |
| 29.1 | 98.543920 | 34.1 | 115.475865 |
| 29.2 | 98.882559 | 34.2 | 115.814504 |
| 29.3 | 99.221198 | 34.3 | 116.153143 |
| 29.4 | 99.559837 | 34.4 | 116.491782 |
| 29.5 | 99.898476 | 34.5 | 116.830421 |
| 29.6 | 100.237114 | 34.6 | 117.169059 |
| 29.7 | 100.575753 | 34.7 | 117.507698 |
| 29.8 | 100.914392 | 34.8 | 117.846337 |
| 29.9 | 101.253031 | 34.9 | 118.184976 |
| | | 35.0 | 118.523615 |

# PRESSURE

## Kilopascals to Inches of Mercury

| kPa | in Hg | kPa | in Hg |
|---|---|---|---|
| 95.0 | 28.053500 | 100.0 | 29.530000 |
| 95.1 | 28.083030 | 100.1 | 29.559530 |
| 95.2 | 28.112560 | 100.2 | 29.589060 |
| 95.3 | 28.142090 | 100.3 | 29.618590 |
| 95.4 | 28.171620 | 100.4 | 29.648120 |
| 95.5 | 28.201150 | 100.5 | 29.677650 |
| 95.6 | 28.230680 | 100.6 | 29.707180 |
| 95.7 | 28.260210 | 100.7 | 29.736710 |
| 95.8 | 28.289740 | 100.8 | 29.766240 |
| 95.9 | 28.319270 | 100.9 | 29.795770 |
| 96.0 | 28.348800 | 101.0 | 29.825300 |
| 96.1 | 28.378330 | 101.1 | 29.854830 |
| 96.2 | 28.407860 | 101.2 | 29.884360 |
| 96.3 | 28.437390 | 101.3 | 29.913890 |
| 96.4 | 28.466920 | 101.4 | 29.943420 |
| 96.5 | 28.496450 | 101.5 | 29.972950 |
| 96.6 | 28.525980 | 101.6 | 30.002480 |
| 96.7 | 28.555510 | 101.7 | 30.032010 |
| 96.8 | 28.585040 | 101.8 | 30.061540 |
| 96.9 | 28.614570 | 101.9 | 30.091070 |
| 97.0 | 28.644100 | 102.0 | 30.120600 |
| 97.1 | 28.673630 | 102.1 | 30.150130 |
| 97.2 | 28.703160 | 102.2 | 30.179660 |
| 97.3 | 28.732690 | 102.3 | 30.209190 |
| 97.4 | 28.762220 | 102.4 | 30.238720 |
| 97.5 | 28.791750 | 102.5 | 30.268250 |
| 97.6 | 28.821280 | 102.6 | 30.297780 |
| 97.7 | 28.850810 | 102.7 | 30.327310 |
| 97.8 | 28.880340 | 102.8 | 30.356840 |
| 97.9 | 28.909870 | 102.9 | 30.386370 |
| 98.0 | 28.939400 | 103.0 | 30.415900 |
| 98.1 | 28.968930 | 103.1 | 30.445430 |
| 98.2 | 28.998460 | 103.2 | 30.474960 |
| 98.3 | 29.027990 | 103.3 | 30.504490 |
| 98.4 | 29.057520 | 103.4 | 30.534020 |
| 98.5 | 29.087050 | 103.5 | 30.563550 |
| 98.6 | 29.116580 | 103.6 | 30.593080 |
| 98.7 | 29.146110 | 103.7 | 30.622610 |
| 98.8 | 29.175640 | 103.8 | 30.652140 |
| 98.9 | 29.205170 | 103.9 | 30.681670 |
| 99.0 | 29.234700 | 104.0 | 30.711200 |
| 99.1 | 29.264230 | 104.1 | 30.740730 |
| 99.2 | 29.293760 | 104.2 | 30.770260 |
| 99.3 | 29.323290 | 104.3 | 30.799790 |
| 99.4 | 29.352820 | 104.4 | 30.829320 |
| 99.5 | 29.382350 | 104.5 | 30.858850 |
| 99.6 | 29.411880 | 104.6 | 30.888380 |
| 99.7 | 29.441410 | 104.7 | 30.917910 |
| 99.8 | 29.470940 | 104.8 | 30.947440 |
| 99.9 | 29.500470 | 104.9 | 30.976970 |
| | | 105.0 | 31.006500 |

## PRESSURE
### Inches of Water to Kilopascals

| in $H_2O$ | kPa | in $H_2O$ | kPa |
|---|---|---|---|
| 1 | .249082 | 51 | 12.703182 |
| 2 | .498164 | 52 | 12.952264 |
| 3 | .747246 | 53 | 13.201346 |
| 4 | .996328 | 54 | 13.450428 |
| 5 | 1.245410 | 55 | 13.699510 |
| 6 | 1.494492 | 56 | 13.948592 |
| 7 | 1.743574 | 57 | 14.197674 |
| 8 | 1.992656 | 58 | 14.446756 |
| 9 | 2.241738 | 59 | 14.695838 |
| 10 | 2.490820 | 60 | 14.944920 |
| 11 | 2.739902 | 61 | 15.194002 |
| 12 | 2.988984 | 62 | 15.443084 |
| 13 | 3.238066 | 63 | 15.692166 |
| 14 | 3.487148 | 64 | 15.941248 |
| 15 | 3.736230 | 65 | 16.190330 |
| 16 | 3.985312 | 66 | 16.439412 |
| 17 | 4.234394 | 67 | 16.688494 |
| 18 | 4.483476 | 68 | 16.937576 |
| 19 | 4.732558 | 69 | 17.186658 |
| 20 | 4.981640 | 70 | 17.435740 |
| 21 | 5.230722 | 71 | 17.684822 |
| 22 | 5.479804 | 72 | 17.933904 |
| 23 | 5.728886 | 73 | 18.182986 |
| 24 | 5.977968 | 74 | 18.432068 |
| 25 | 6.227050 | 75 | 18.681150 |
| 26 | 6.476132 | 76 | 18.930232 |
| 27 | 6.725214 | 77 | 19.179314 |
| 28 | 6.974296 | 78 | 19.428396 |
| 29 | 7.223378 | 79 | 19.677478 |
| 30 | 7.472460 | 80 | 19.926560 |
| 31 | 7.721542 | 81 | 20.175642 |
| 32 | 7.970624 | 82 | 20.424724 |
| 33 | 8.219706 | 83 | 20.673806 |
| 34 | 8.468788 | 84 | 20.922888 |
| 35 | 8.717870 | 85 | 21.171970 |
| 36 | 8.966952 | 86 | 21.421052 |
| 37 | 9.216034 | 87 | 21.670134 |
| 38 | 9.465116 | 88 | 21.919216 |
| 39 | 9.714198 | 89 | 22.168298 |
| 40 | 9.963280 | 90 | 22.417380 |
| 41 | 10.212362 | 91 | 22.666462 |
| 42 | 10.461444 | 92 | 22.915544 |
| 43 | 10.710526 | 93 | 23.164626 |
| 44 | 10.959608 | 94 | 23.413708 |
| 45 | 11.208690 | 95 | 23.662790 |
| 46 | 11.457772 | 96 | 23.911872 |
| 47 | 11.706854 | 97 | 24.160954 |
| 48 | 11.955936 | 98 | 24.410036 |
| 49 | 12.205018 | 99 | 24.659118 |
| 50 | 12.454100 | 100 | 24.908200 |

## PRESSURE
### Kilopascals to Inches of Water

| kPa | in $H_2O$ | kPa | in $H_2O$ |
|---|---|---|---|
| 1 | 4.014744 | 51 | 204.751944 |
| 2 | 8.029488 | 52 | 208.766688 |
| 3 | 12.044232 | 53 | 212.781432 |
| 4 | 16.058976 | 54 | 216.796176 |
| 5 | 20.073720 | 55 | 220.810920 |
| 6 | 24.088464 | 56 | 224.825664 |
| 7 | 28.103208 | 57 | 228.840408 |
| 8 | 32.117952 | 58 | 232.855152 |
| 9 | 36.132696 | 59 | 236.869896 |
| 10 | 40.147440 | 60 | 240.884640 |
| 11 | 44.162184 | 61 | 244.899384 |
| 12 | 48.176928 | 62 | 248.914128 |
| 13 | 52.191672 | 63 | 252.928872 |
| 14 | 56.206416 | 64 | 256.943616 |
| 15 | 60.221160 | 65 | 260.958360 |
| 16 | 64.235904 | 66 | 264.973104 |
| 17 | 68.250648 | 67 | 268.987848 |
| 18 | 72.265392 | 68 | 273.002592 |
| 19 | 76.280136 | 69 | 277.017336 |
| 20 | 80.294880 | 70 | 281.032080 |
| 21 | 84.309624 | 71 | 285.046824 |
| 22 | 88.324368 | 72 | 289.061568 |
| 23 | 92.339112 | 73 | 293.076312 |
| 24 | 96.353856 | 74 | 297.091056 |
| 25 | 100.368600 | 75 | 301.105800 |
| 26 | 104.383344 | 76 | 305.120544 |
| 27 | 108.398088 | 77 | 309.135288 |
| 28 | 112.412832 | 78 | 313.150032 |
| 29 | 116.427576 | 79 | 317.164776 |
| 30 | 120.442320 | 80 | 321.179520 |
| 31 | 124.457064 | 81 | 325.194264 |
| 32 | 128.471808 | 82 | 329.209008 |
| 33 | 132.486552 | 83 | 333.223752 |
| 34 | 136.501296 | 84 | 337.238496 |
| 35 | 140.516040 | 85 | 341.253240 |
| 36 | 144.530784 | 86 | 345.267984 |
| 37 | 148.545528 | 87 | 349.282728 |
| 38 | 152.560272 | 88 | 353.297472 |
| 39 | 156.575016 | 89 | 357.312216 |
| 40 | 160.589760 | 90 | 361.326960 |
| 41 | 164.604504 | 91 | 365.341704 |
| 42 | 168.619248 | 92 | 369.356448 |
| 43 | 172.633992 | 93 | 373.371192 |
| 44 | 176.648736 | 94 | 377.385936 |
| 45 | 180.663480 | 95 | 381.400680 |
| 46 | 184.678224 | 96 | 385.415424 |
| 47 | 188.692968 | 97 | 389.430168 |
| 48 | 192.707712 | 98 | 393.444912 |
| 49 | 196.722456 | 99 | 397.459656 |
| 50 | 200.737200 | 100 | 401.474400 |

## ENERGY, WORK
### Foot-Pounds to Joules

| ft·lb | J | ft·lb | J |
|---|---|---|---|
| 1 | 1.355818 | 51 | 69.146718 |
| 2 | 2.711636 | 52 | 70.502536 |
| 3 | 4.067454 | 53 | 71.858354 |
| 4 | 5.423272 | 54 | 73.214172 |
| 5 | 6.779090 | 55 | 74.569990 |
| 6 | 8.134908 | 56 | 75.925808 |
| 7 | 9.490726 | 57 | 77.281626 |
| 8 | 10.846544 | 58 | 78.637444 |
| 9 | 12.202362 | 59 | 79.993262 |
| 10 | 13.558180 | 60 | 81.349080 |
| 11 | 14.913998 | 61 | 82.704898 |
| 12 | 16.269816 | 62 | 84.060716 |
| 13 | 17.625634 | 63 | 85.416534 |
| 14 | 18.981452 | 64 | 86.772352 |
| 15 | 20.337270 | 65 | 88.128170 |
| 16 | 21.693088 | 66 | 89.483988 |
| 17 | 23.048906 | 67 | 90.839806 |
| 18 | 24.404724 | 68 | 92.195624 |
| 19 | 25.760542 | 69 | 93.551442 |
| 20 | 27.116360 | 70 | 94.907260 |
| 21 | 28.472178 | 71 | 96.263078 |
| 22 | 29.827996 | 72 | 97.618896 |
| 23 | 31.183814 | 73 | 98.974714 |
| 24 | 32.539632 | 74 | 100.330532 |
| 25 | 33.895450 | 75 | 101.686350 |
| 26 | 35.251268 | 76 | 103.042168 |
| 27 | 36.607086 | 77 | 104.397986 |
| 28 | 37.962904 | 78 | 105.753804 |
| 29 | 39.318722 | 79 | 107.109622 |
| 30 | 40.674540 | 80 | 108.465440 |
| 31 | 42.030358 | 81 | 109.821258 |
| 32 | 43.386176 | 82 | 111.177076 |
| 33 | 44.741994 | 83 | 112.532894 |
| 34 | 46.097812 | 84 | 113.888712 |
| 35 | 47.453630 | 85 | 115.244530 |
| 36 | 48.809448 | 86 | 116.600348 |
| 37 | 50.165266 | 87 | 117.956166 |
| 38 | 51.521084 | 88 | 119.311984 |
| 39 | 52.876902 | 89 | 120.667802 |
| 40 | 54.232720 | 90 | 122.023620 |
| 41 | 55.588538 | 91 | 123.379438 |
| 42 | 56.944356 | 92 | 124.735256 |
| 43 | 58.300174 | 93 | 126.091074 |
| 44 | 59.655992 | 94 | 127.446892 |
| 45 | 61.011810 | 95 | 128.802710 |
| 46 | 62.367628 | 96 | 130.158528 |
| 47 | 63.723446 | 97 | 131.514346 |
| 48 | 65.079264 | 98 | 132.870164 |
| 49 | 66.435082 | 99 | 134.225982 |
| 50 | 67.790900 | 100 | 135.581800 |

## ENERGY, WORK
### Joules to Foot-Pounds

| J | ft·lb | J | ft·lb |
|---|---|---|---|
| 1 | .737562 | 51 | 37.615662 |
| 2 | 1.475124 | 52 | 38.353224 |
| 3 | 2.212686 | 53 | 39.090786 |
| 4 | 2.950248 | 54 | 39.828348 |
| 5 | 3.687810 | 55 | 40.565910 |
| 6 | 4.425372 | 56 | 41.303472 |
| 7 | 5.162934 | 57 | 42.041034 |
| 8 | 5.900496 | 58 | 42.778596 |
| 9 | 6.638058 | 59 | 43.516158 |
| 10 | 7.375620 | 60 | 44.253720 |
| 11 | 8.113182 | 61 | 44.991282 |
| 12 | 8.850744 | 62 | 45.728844 |
| 13 | 9.588306 | 63 | 46.466406 |
| 14 | 10.325868 | 64 | 47.203968 |
| 15 | 11.063430 | 65 | 47.941530 |
| 16 | 11.800992 | 66 | 48.679092 |
| 17 | 12.538554 | 67 | 49.416654 |
| 18 | 13.276116 | 68 | 50.154216 |
| 19 | 14.013678 | 69 | 50.891778 |
| 20 | 14.751240 | 70 | 51.629340 |
| 21 | 15.488802 | 71 | 52.366902 |
| 22 | 16.226364 | 72 | 53.104464 |
| 23 | 16.963926 | 73 | 53.842026 |
| 24 | 17.701488 | 74 | 54.579588 |
| 25 | 18.439050 | 75 | 55.317150 |
| 26 | 19.176612 | 76 | 56.054712 |
| 27 | 19.914174 | 77 | 56.792274 |
| 28 | 20.651736 | 78 | 57.529836 |
| 29 | 21.389298 | 79 | 58.267398 |
| 30 | 22.126860 | 80 | 59.004960 |
| 31 | 22.864422 | 81 | 59.742522 |
| 32 | 23.601984 | 82 | 60.480084 |
| 33 | 24.339546 | 83 | 61.217646 |
| 34 | 25.077108 | 84 | 61.955208 |
| 35 | 25.814670 | 85 | 62.692770 |
| 36 | 26.552232 | 86 | 63.430332 |
| 37 | 27.289794 | 87 | 64.167894 |
| 38 | 28.027356 | 88 | 64.905456 |
| 39 | 28.764918 | 89 | 65.643018 |
| 40 | 29.502480 | 90 | 66.380580 |
| 41 | 30.240042 | 91 | 67.118142 |
| 42 | 30.977604 | 92 | 67.855704 |
| 43 | 31.715166 | 93 | 68.593266 |
| 44 | 32.452728 | 94 | 69.330828 |
| 45 | 33.190290 | 95 | 70.068390 |
| 46 | 33.927852 | 96 | 70.805952 |
| 47 | 34.665414 | 97 | 71.543514 |
| 48 | 35.402976 | 98 | 72.281076 |
| 49 | 36.140538 | 99 | 73.018638 |
| 50 | 36.878100 | 100 | 73.756200 |

## ENERGY, WORK
### Meter-Kilograms to Joules

| m·kg | J | m·kg | J |
|---|---|---|---|
| 1 | 9.806650 | 51 | 500.139150 |
| 2 | 19.613300 | 52 | 509.945800 |
| 3 | 29.419950 | 53 | 519.752450 |
| 4 | 39.226600 | 54 | 529.559100 |
| 5 | 49.033250 | 55 | 539.365750 |
| 6 | 58.839900 | 56 | 549.172400 |
| 7 | 68.646550 | 57 | 558.979050 |
| 8 | 78.453200 | 58 | 568.785700 |
| 9 | 88.259850 | 59 | 578.592350 |
| 10 | 98.066500 | 60 | 588.399000 |
| 11 | 107.873150 | 61 | 598.205650 |
| 12 | 117.679800 | 62 | 608.012300 |
| 13 | 127.486450 | 63 | 617.818950 |
| 14 | 137.293100 | 64 | 627.625600 |
| 15 | 147.099750 | 65 | 637.432250 |
| 16 | 156.906400 | 66 | 647.238900 |
| 17 | 166.713050 | 67 | 657.045550 |
| 18 | 176.519700 | 68 | 666.852200 |
| 19 | 186.326350 | 69 | 676.658850 |
| 20 | 196.133000 | 70 | 686.465500 |
| 21 | 205.939650 | 71 | 696.272150 |
| 22 | 215.746300 | 72 | 706.078800 |
| 23 | 225.552950 | 73 | 715.885450 |
| 24 | 235.359600 | 74 | 725.692100 |
| 25 | 245.166250 | 75 | 735.498750 |
| 26 | 254.972900 | 76 | 745.305400 |
| 27 | 264.779550 | 77 | 755.112050 |
| 28 | 274.586200 | 78 | 764.918700 |
| 29 | 284.392850 | 79 | 774.725350 |
| 30 | 294.199500 | 80 | 784.532000 |
| 31 | 304.006150 | 81 | 794.338650 |
| 32 | 313.812800 | 82 | 804.145300 |
| 33 | 323.619450 | 83 | 813.951950 |
| 34 | 333.426100 | 84 | 823.758600 |
| 35 | 343.232750 | 85 | 833.565250 |
| 36 | 353.039400 | 86 | 843.371900 |
| 37 | 362.846050 | 87 | 853.178550 |
| 38 | 372.652700 | 88 | 862.985200 |
| 39 | 382.459350 | 89 | 872.791850 |
| 40 | 392.266000 | 90 | 882.598500 |
| 41 | 402.072650 | 91 | 892.405150 |
| 42 | 411.879300 | 92 | 902.211800 |
| 43 | 421.685950 | 93 | 912.018450 |
| 44 | 431.492600 | 94 | 921.825100 |
| 45 | 441.299250 | 95 | 931.631750 |
| 46 | 451.105900 | 96 | 941.438400 |
| 47 | 460.912550 | 97 | 951.245050 |
| 48 | 470.719200 | 98 | 961.051700 |
| 49 | 480.525850 | 99 | 970.858350 |
| 50 | 490.332500 | 100 | 980.665000 |

## ENERGY, WORK
### Joules to Meter-Kilograms

| J | m·kg | J | m·kg |
|---|---|---|---|
| 1 | .101972 | 51 | 5.200572 |
| 2 | .203944 | 52 | 5.302544 |
| 3 | .305916 | 53 | 5.404516 |
| 4 | .407888 | 54 | 5.506488 |
| 5 | .509860 | 55 | 5.608460 |
| 6 | .611832 | 56 | 5.710432 |
| 7 | .713804 | 57 | 5.812404 |
| 8 | .815776 | 58 | 5.914376 |
| 9 | .917748 | 59 | 6.016348 |
| 10 | 1.019720 | 60 | 6.118320 |
| 11 | 1.121692 | 61 | 6.220292 |
| 12 | 1.223664 | 62 | 6.322264 |
| 13 | 1.325636 | 63 | 6.424236 |
| 14 | 1.427608 | 64 | 6.526208 |
| 15 | 1.529580 | 65 | 6.628180 |
| 16 | 1.631552 | 66 | 6.730152 |
| 17 | 1.733524 | 67 | 6.832124 |
| 18 | 1.835496 | 68 | 6.934096 |
| 19 | 1.937468 | 69 | 7.036068 |
| 20 | 2.039440 | 70 | 7.138040 |
| 21 | 2.141412 | 71 | 7.240012 |
| 22 | 2.243384 | 72 | 7.341984 |
| 23 | 2.345356 | 73 | 7.443956 |
| 24 | 2.447328 | 74 | 7.545928 |
| 25 | 2.549300 | 75 | 7.647900 |
| 26 | 2.651272 | 76 | 7.749872 |
| 27 | 2.753244 | 77 | 7.851844 |
| 28 | 2.855216 | 78 | 7.953816 |
| 29 | 2.957188 | 79 | 8.055788 |
| 30 | 3.059160 | 80 | 8.157760 |
| 31 | 3.161132 | 81 | 8.259732 |
| 32 | 3.263104 | 82 | 8.361704 |
| 33 | 3.365076 | 83 | 8.463676 |
| 34 | 3.467048 | 84 | 8.565648 |
| 35 | 3.569020 | 85 | 8.667620 |
| 36 | 3.670992 | 86 | 8.769592 |
| 37 | 3.772964 | 87 | 8.871564 |
| 38 | 3.874936 | 88 | 8.973536 |
| 39 | 3.976908 | 89 | 9.075508 |
| 40 | 4.078880 | 90 | 9.177480 |
| 41 | 4.180852 | 91 | 9.279452 |
| 42 | 4.282824 | 92 | 9.381424 |
| 43 | 4.384796 | 93 | 9.483396 |
| 44 | 4.486768 | 94 | 9.585368 |
| 45 | 4.588740 | 95 | 9.687340 |
| 46 | 4.690712 | 96 | 9.789312 |
| 47 | 4.792684 | 97 | 9.891284 |
| 48 | 4.894656 | 98 | 9.993256 |
| 49 | 4.996628 | 99 | 10.095228 |
| 50 | 5.098600 | 100 | 10.197200 |

## ENERGY, WORK
### British Thermal Units to Kilojoules

| Btu | kJ | Btu | kJ |
|---|---|---|---|
| 1 | 1.054800 | 51 | 53.794800 |
| 2 | 2.109600 | 52 | 54.849600 |
| 3 | 3.164400 | 53 | 55.904400 |
| 4 | 4.219200 | 54 | 56.959200 |
| 5 | 5.274000 | 55 | 58.014000 |
| 6 | 6.328800 | 56 | 59.068800 |
| 7 | 7.383600 | 57 | 60.123600 |
| 8 | 8.438400 | 58 | 61.178400 |
| 9 | 9.493200 | 59 | 62.233200 |
| 10 | 10.548000 | 60 | 63.288000 |
| 11 | 11.602800 | 61 | 64.342800 |
| 12 | 12.657600 | 62 | 65.397600 |
| 13 | 13.712400 | 63 | 66.452400 |
| 14 | 14.767200 | 64 | 67.507200 |
| 15 | 15.822000 | 65 | 68.562000 |
| 16 | 16.876800 | 66 | 69.616800 |
| 17 | 17.931600 | 67 | 70.671600 |
| 18 | 18.986400 | 68 | 71.726400 |
| 19 | 20.041200 | 69 | 72.781200 |
| 20 | 21.096000 | 70 | 73.836000 |
| 21 | 22.150800 | 71 | 74.890800 |
| 22 | 23.205600 | 72 | 75.945600 |
| 23 | 24.260400 | 73 | 77.000400 |
| 24 | 25.315200 | 74 | 78.055200 |
| 25 | 26.370000 | 75 | 79.110000 |
| 26 | 27.424800 | 76 | 80.164800 |
| 27 | 28.479600 | 77 | 81.219600 |
| 28 | 29.534400 | 78 | 82.274400 |
| 29 | 30.589200 | 79 | 83.329200 |
| 30 | 31.644000 | 80 | 84.384000 |
| 31 | 32.698800 | 81 | 85.438800 |
| 32 | 33.753600 | 82 | 86.493600 |
| 33 | 34.808400 | 83 | 87.548400 |
| 34 | 35.863200 | 84 | 88.603200 |
| 35 | 36.918000 | 85 | 89.658000 |
| 36 | 37.972800 | 86 | 90.712800 |
| 37 | 39.027600 | 87 | 91.767600 |
| 38 | 40.082400 | 88 | 92.822400 |
| 39 | 41.137200 | 89 | 93.877200 |
| 40 | 42.192000 | 90 | 94.932000 |
| 41 | 43.246800 | 91 | 95.986800 |
| 42 | 44.301600 | 92 | 97.041600 |
| 43 | 45.356400 | 93 | 98.096400 |
| 44 | 46.411200 | 94 | 99.151200 |
| 45 | 47.466000 | 95 | 100.206000 |
| 46 | 48.520800 | 96 | 101.260800 |
| 47 | 49.575600 | 97 | 102.315600 |
| 48 | 50.630400 | 98 | 103.370400 |
| 49 | 51.685200 | 99 | 104.425200 |
| 50 | 52.740000 | 100 | 105.480000 |

## ENERGY, WORK
### Kilojoules to British Thermal Units

| kJ | Btu | kJ | Btu |
|---|---|---|---|
| 1 | .948047 | 51 | 48.350397 |
| 2 | 1.896094 | 52 | 49.298444 |
| 3 | 2.844141 | 53 | 50.246491 |
| 4 | 3.792188 | 54 | 51.194538 |
| 5 | 4.740235 | 55 | 52.142585 |
| 6 | 5.688282 | 56 | 53.090632 |
| 7 | 6.636329 | 57 | 54.038679 |
| 8 | 7.584376 | 58 | 54.986726 |
| 9 | 8.532423 | 59 | 55.934773 |
| 10 | 9.480470 | 60 | 56.882820 |
| 11 | 10.428517 | 61 | 57.830867 |
| 12 | 11.376564 | 62 | 58.778914 |
| 13 | 12.324611 | 63 | 59.726961 |
| 14 | 13.272658 | 64 | 60.675008 |
| 15 | 14.220705 | 65 | 61.623055 |
| 16 | 15.168752 | 66 | 62.571102 |
| 17 | 16.116799 | 67 | 63.519149 |
| 18 | 17.064846 | 68 | 64.467196 |
| 19 | 18.012893 | 69 | 65.415243 |
| 20 | 18.960940 | 70 | 66.363290 |
| 21 | 19.908987 | 71 | 67.311337 |
| 22 | 20.857034 | 72 | 68.259384 |
| 23 | 21.805081 | 73 | 69.207431 |
| 24 | 22.753128 | 74 | 70.155478 |
| 25 | 23.701175 | 75 | 71.103525 |
| 26 | 24.649222 | 76 | 72.051572 |
| 27 | 25.597269 | 77 | 72.999619 |
| 28 | 26.545316 | 78 | 73.947666 |
| 29 | 27.493363 | 79 | 74.895713 |
| 30 | 28.441410 | 80 | 75.843760 |
| 31 | 29.389457 | 81 | 76.791807 |
| 32 | 30.337504 | 82 | 77.739854 |
| 33 | 31.285551 | 83 | 78.687901 |
| 34 | 32.233598 | 84 | 79.635948 |
| 35 | 33.181645 | 85 | 80.583995 |
| 36 | 34.129692 | 86 | 81.532042 |
| 37 | 35.077739 | 87 | 82.480089 |
| 38 | 36.025786 | 88 | 83.428136 |
| 39 | 36.973833 | 89 | 84.376183 |
| 40 | 37.921880 | 90 | 85.324230 |
| 41 | 38.869927 | 91 | 86.272277 |
| 42 | 39.817974 | 92 | 87.220324 |
| 43 | 40.766021 | 93 | 88.168371 |
| 44 | 41.714068 | 94 | 89.116418 |
| 45 | 42.662115 | 95 | 90.064465 |
| 46 | 43.610162 | 96 | 91.012512 |
| 47 | 44.558209 | 97 | 91.960559 |
| 48 | 45.506256 | 98 | 92.908606 |
| 49 | 46.454303 | 99 | 93.856653 |
| 50 | 47.402350 | 100 | 94.804700 |

## ENERGY, WORK
### Calories (tech) to Joules

| cal | J | cal | J |
|---|---|---|---|
| 1 | 4.186501 | 51 | 213.511551 |
| 2 | 8.373002 | 52 | 217.698052 |
| 3 | 12.559503 | 53 | 221.884553 |
| 4 | 16.746004 | 54 | 226.071054 |
| 5 | 20.932505 | 55 | 230.257555 |
| 6 | 25.119006 | 56 | 234.444056 |
| 7 | 29.305507 | 57 | 238.630557 |
| 8 | 33.492008 | 58 | 242.817058 |
| 9 | 37.678509 | 59 | 247.003559 |
| 10 | 41.865010 | 60 | 251.190060 |
| 11 | 46.051511 | 61 | 255.376561 |
| 12 | 50.238012 | 62 | 259.563062 |
| 13 | 54.424513 | 63 | 263.749563 |
| 14 | 58.611014 | 64 | 267.936064 |
| 15 | 62.797515 | 65 | 272.122565 |
| 16 | 66.984016 | 66 | 276.309066 |
| 17 | 71.170517 | 67 | 280.495567 |
| 18 | 75.357018 | 68 | 284.682068 |
| 19 | 79.543519 | 69 | 288.868569 |
| 20 | 83.730020 | 70 | 293.055070 |
| 21 | 87.916521 | 71 | 297.241571 |
| 22 | 92.103022 | 72 | 301.428072 |
| 23 | 96.289523 | 73 | 305.614573 |
| 24 | 100.476024 | 74 | 309.801074 |
| 25 | 104.662525 | 75 | 313.987575 |
| 26 | 108.849026 | 76 | 318.174076 |
| 27 | 113.035527 | 77 | 322.360577 |
| 28 | 117.222028 | 78 | 326.547078 |
| 29 | 121.408529 | 79 | 330.733579 |
| 30 | 125.595030 | 80 | 334.920080 |
| 31 | 129.781531 | 81 | 339.106581 |
| 32 | 133.968032 | 82 | 343.293082 |
| 33 | 138.154533 | 83 | 347.479583 |
| 34 | 142.341034 | 84 | 351.666084 |
| 35 | 146.527535 | 85 | 355.852585 |
| 36 | 150.714036 | 86 | 360.039086 |
| 37 | 154.900537 | 87 | 364.225587 |
| 38 | 159.087038 | 88 | 368.412088 |
| 39 | 163.273539 | 89 | 372.598589 |
| 40 | 167.460040 | 90 | 376.785090 |
| 41 | 171.646541 | 91 | 380.971591 |
| 42 | 175.833042 | 92 | 385.158092 |
| 43 | 180.019543 | 93 | 389.344593 |
| 44 | 184.206044 | 94 | 393.531094 |
| 45 | 188.392545 | 95 | 397.717595 |
| 46 | 192.579046 | 96 | 401.904096 |
| 47 | 196.765547 | 97 | 406.090597 |
| 48 | 200.952048 | 98 | 410.277098 |
| 49 | 205.138549 | 99 | 414.463599 |
| 50 | 209.325050 | 100 | 418.650100 |

## ENERGY, WORK
### Joules to Calories (tech)

| J | cal | J | cal |
|---|---|---|---|
| 1 | .238863 | 51 | 12.182013 |
| 2 | .477726 | 52 | 12.420876 |
| 3 | .716589 | 53 | 12.659739 |
| 4 | .955452 | 54 | 12.898602 |
| 5 | 1.194315 | 55 | 13.137465 |
| 6 | 1.433178 | 56 | 13.376328 |
| 7 | 1.672041 | 57 | 13.615191 |
| 8 | 1.910904 | 58 | 13.854054 |
| 9 | 2.149767 | 59 | 14.092917 |
| 10 | 2.388630 | 60 | 14.331780 |
| 11 | 2.627493 | 61 | 14.570643 |
| 12 | 2.866356 | 62 | 14.809506 |
| 13 | 3.105219 | 63 | 15.048369 |
| 14 | 3.344082 | 64 | 15.287232 |
| 15 | 3.582945 | 65 | 15.526095 |
| 16 | 3.821808 | 66 | 15.764958 |
| 17 | 4.060671 | 67 | 16.003821 |
| 18 | 4.299534 | 68 | 16.242684 |
| 19 | 4.538397 | 69 | 16.481547 |
| 20 | 4.777260 | 70 | 16.720410 |
| 21 | 5.016123 | 71 | 16.959273 |
| 22 | 5.254986 | 72 | 17.198136 |
| 23 | 5.493849 | 73 | 17.436999 |
| 24 | 5.732712 | 74 | 17.675862 |
| 25 | 5.971575 | 75 | 17.914725 |
| 26 | 6.210438 | 76 | 18.153588 |
| 27 | 6.449301 | 77 | 18.392451 |
| 28 | 6.688164 | 78 | 18.631314 |
| 29 | 6.927027 | 79 | 18.870177 |
| 30 | 7.165890 | 80 | 19.109040 |
| 31 | 7.404753 | 81 | 19.347903 |
| 32 | 7.643616 | 82 | 19.586766 |
| 33 | 7.882479 | 83 | 19.825629 |
| 34 | 8.121342 | 84 | 20.064492 |
| 35 | 8.360205 | 85 | 20.303355 |
| 36 | 8.599068 | 86 | 20.542218 |
| 37 | 8.837931 | 87 | 20.781081 |
| 38 | 9.076794 | 88 | 21.019944 |
| 39 | 9.315657 | 89 | 21.258807 |
| 40 | 9.554520 | 90 | 21.497670 |
| 41 | 9.793383 | 91 | 21.736533 |
| 42 | 10.032246 | 92 | 21.975396 |
| 43 | 10.271109 | 93 | 22.214259 |
| 44 | 10.509972 | 94 | 22.453122 |
| 45 | 10.748835 | 95 | 22.691985 |
| 46 | 10.987698 | 96 | 22.930848 |
| 47 | 11.226561 | 97 | 23.169711 |
| 48 | 11.465424 | 98 | 23.408574 |
| 49 | 11.704287 | 99 | 23.647437 |
| 50 | 11.943150 | 100 | 23.886300 |

## ENERGY, WORK
### Kilowatt-hours to Megajoules

| kWh | MJ | kWh | MJ |
|---|---|---|---|
| 1 | 3.600000 | 51 | 183.600000 |
| 2 | 7.200000 | 52 | 187.200000 |
| 3 | 10.800000 | 53 | 190.800000 |
| 4 | 14.400000 | 54 | 194.400000 |
| 5 | 18.000000 | 55 | 198.000000 |
| 6 | 21.600000 | 56 | 201.600000 |
| 7 | 25.200000 | 57 | 205.200000 |
| 8 | 28.800000 | 58 | 208.800000 |
| 9 | 32.400000 | 59 | 212.400000 |
| 10 | 36.000000 | 60 | 216.000000 |
| 11 | 39.600000 | 61 | 219.600000 |
| 12 | 43.200000 | 62 | 223.200000 |
| 13 | 46.800000 | 63 | 226.800000 |
| 14 | 50.400000 | 64 | 230.400000 |
| 15 | 54.000000 | 65 | 234.000000 |
| 16 | 57.600000 | 66 | 237.600000 |
| 17 | 61.200000 | 67 | 241.200000 |
| 18 | 64.800000 | 68 | 244.800000 |
| 19 | 68.400000 | 69 | 248.400000 |
| 20 | 72.000000 | 70 | 252.000000 |
| 21 | 75.600000 | 71 | 255.600000 |
| 22 | 79.200000 | 72 | 259.200000 |
| 23 | 82.800000 | 73 | 262.800000 |
| 24 | 86.400000 | 74 | 266.400000 |
| 25 | 90.000000 | 75 | 270.000000 |
| 26 | 93.600000 | 76 | 273.600000 |
| 27 | 97.200000 | 77 | 277.200000 |
| 28 | 100.800000 | 78 | 280.800000 |
| 29 | 104.400000 | 79 | 284.400000 |
| 30 | 108.000000 | 80 | 288.000000 |
| 31 | 111.600000 | 81 | 291.600000 |
| 32 | 115.200000 | 82 | 295.200000 |
| 33 | 118.800000 | 83 | 298.800000 |
| 34 | 122.400000 | 84 | 302.400000 |
| 35 | 126.000000 | 85 | 306.000000 |
| 36 | 129.600000 | 86 | 309.600000 |
| 37 | 133.200000 | 87 | 313.200000 |
| 38 | 136.800000 | 88 | 316.800000 |
| 39 | 140.400000 | 89 | 320.400000 |
| 40 | 144.000000 | 90 | 324.000000 |
| 41 | 147.600000 | 91 | 327.600000 |
| 42 | 151.200000 | 92 | 331.200000 |
| 43 | 154.800000 | 93 | 334.800000 |
| 44 | 158.400000 | 94 | 338.400000 |
| 45 | 162.000000 | 95 | 342.000000 |
| 46 | 165.600000 | 96 | 345.600000 |
| 47 | 169.200000 | 97 | 349.200000 |
| 48 | 172.800000 | 98 | 352.800000 |
| 49 | 176.400000 | 99 | 356.400000 |
| 50 | 180.000000 | 100 | 360.000000 |

## ENERGY, WORK
### Megajoules to Kilowatt-hours

| MJ | kWh | MJ | kWh |
|---|---|---|---|
| 1 | .277777 | 51 | 14.166627 |
| 2 | .555554 | 52 | 14.444404 |
| 3 | .833331 | 53 | 14.722181 |
| 4 | 1.111108 | 54 | 14.999958 |
| 5 | 1.388885 | 55 | 15.277735 |
| 6 | 1.666662 | 56 | 15.555512 |
| 7 | 1.944439 | 57 | 15.833289 |
| 8 | 2.222216 | 58 | 16.111066 |
| 9 | 2.499993 | 59 | 16.388843 |
| 10 | 2.777770 | 60 | 16.666620 |
| 11 | 3.055547 | 61 | 16.944397 |
| 12 | 3.333324 | 62 | 17.222174 |
| 13 | 3.611101 | 63 | 17.499951 |
| 14 | 3.888878 | 64 | 17.777728 |
| 15 | 4.166655 | 65 | 18.055505 |
| 16 | 4.444432 | 66 | 18.333282 |
| 17 | 4.722209 | 67 | 18.611059 |
| 18 | 4.999986 | 68 | 18.888836 |
| 19 | 5.277763 | 69 | 19.166613 |
| 20 | 5.555540 | 70 | 19.444390 |
| 21 | 5.833317 | 71 | 19.722167 |
| 22 | 6.111094 | 72 | 19.999944 |
| 23 | 6.388871 | 73 | 20.277721 |
| 24 | 6.666648 | 74 | 20.555498 |
| 25 | 6.944425 | 75 | 20.833275 |
| 26 | 7.222202 | 76 | 21.111052 |
| 27 | 7.499979 | 77 | 21.388829 |
| 28 | 7.777756 | 78 | 21.666606 |
| 29 | 8.055533 | 79 | 21.944383 |
| 30 | 8.333310 | 80 | 22.222160 |
| 31 | 8.611087 | 81 | 22.499937 |
| 32 | 8.888864 | 82 | 22.777714 |
| 33 | 9.166641 | 83 | 23.055491 |
| 34 | 9.444418 | 84 | 23.333268 |
| 35 | 9.722195 | 85 | 23.611045 |
| 36 | 9.999972 | 86 | 23.888822 |
| 37 | 10.277749 | 87 | 24.166599 |
| 38 | 10.555526 | 88 | 24.444376 |
| 39 | 10.833303 | 89 | 24.722153 |
| 40 | 11.111080 | 90 | 24.999930 |
| 41 | 11.388857 | 91 | 25.277707 |
| 42 | 11.666634 | 92 | 25.555484 |
| 43 | 11.944411 | 93 | 25.833261 |
| 44 | 12.222188 | 94 | 26.111038 |
| 45 | 12.499965 | 95 | 26.388815 |
| 46 | 12.777742 | 96 | 26.666592 |
| 47 | 13.055519 | 97 | 26.944369 |
| 48 | 13.333296 | 98 | 27.222146 |
| 49 | 13.611073 | 99 | 27.499923 |
| 50 | 13.888850 | 100 | 27.777700 |

## POWER
### Foot-pounds per Second to Watts

| ft·lb/s | W | ft·lb/s | W |
|---|---|---|---|
| 1 | 1.355818 | 51 | 69.146718 |
| 2 | 2.711636 | 52 | 70.502536 |
| 3 | 4.067454 | 53 | 71.858354 |
| 4 | 5.423272 | 54 | 73.214172 |
| 5 | 6.779090 | 55 | 74.569990 |
| 6 | 8.134908 | 56 | 75.925808 |
| 7 | 9.490726 | 57 | 77.281626 |
| 8 | 10.846544 | 58 | 78.637444 |
| 9 | 12.202362 | 59 | 79.993262 |
| 10 | 13.558180 | 60 | 81.349080 |
| 11 | 14.913998 | 61 | 82.704898 |
| 12 | 16.269816 | 62 | 84.060716 |
| 13 | 17.625634 | 63 | 85.416534 |
| 14 | 18.981452 | 64 | 86.772352 |
| 15 | 20.337270 | 65 | 88.128170 |
| 16 | 21.693088 | 66 | 89.483988 |
| 17 | 23.048906 | 67 | 90.839806 |
| 18 | 24.404724 | 68 | 92.195624 |
| 19 | 25.760542 | 69 | 93.551442 |
| 20 | 27.116360 | 70 | 94.907260 |
| 21 | 28.472178 | 71 | 96.263078 |
| 22 | 29.827996 | 72 | 97.618896 |
| 23 | 31.183814 | 73 | 98.974714 |
| 24 | 32.539632 | 74 | 100.330532 |
| 25 | 33.895450 | 75 | 101.686350 |
| 26 | 35.251268 | 76 | 103.042168 |
| 27 | 36.607086 | 77 | 104.397986 |
| 28 | 37.962904 | 78 | 105.753804 |
| 29 | 39.318722 | 79 | 107.109622 |
| 30 | 40.674540 | 80 | 108.465440 |
| 31 | 42.030358 | 81 | 109.821258 |
| 32 | 43.386176 | 82 | 111.177076 |
| 33 | 44.741994 | 83 | 112.532894 |
| 34 | 46.097812 | 84 | 113.888712 |
| 35 | 47.453630 | 85 | 115.244530 |
| 36 | 48.809448 | 86 | 116.600348 |
| 37 | 50.165266 | 87 | 117.956166 |
| 38 | 51.521084 | 88 | 119.311984 |
| 39 | 52.876902 | 89 | 120.667802 |
| 40 | 54.232720 | 90 | 122.023620 |
| 41 | 55.588538 | 91 | 123.379438 |
| 42 | 56.944356 | 92 | 124.735256 |
| 43 | 58.300174 | 93 | 126.091074 |
| 44 | 59.655992 | 94 | 127.446892 |
| 45 | 61.011810 | 95 | 128.802710 |
| 46 | 62.367628 | 96 | 130.158528 |
| 47 | 63.723446 | 97 | 131.514346 |
| 48 | 65.079264 | 98 | 132.870164 |
| 49 | 66.435082 | 99 | 134.225982 |
| 50 | 67.790900 | 100 | 135.581800 |

## POWER
### Watts to Foot-pounds per Second

| W | ft·lb/s | W | ft·lb/s |
|---|---|---|---|
| 1 | .737562 | 51 | 37.615662 |
| 2 | 1.475124 | 52 | 38.353224 |
| 3 | 2.212686 | 53 | 39.090786 |
| 4 | 2.950248 | 54 | 39.828348 |
| 5 | 3.687810 | 55 | 40.565910 |
| 6 | 4.425372 | 56 | 41.303472 |
| 7 | 5.162934 | 57 | 42.041034 |
| 8 | 5.900496 | 58 | 42.778596 |
| 9 | 6.638058 | 59 | 43.516158 |
| 10 | 7.375620 | 60 | 44.253720 |
| 11 | 8.113182 | 61 | 44.991282 |
| 12 | 8.850744 | 62 | 45.728844 |
| 13 | 9.588306 | 63 | 46.466406 |
| 14 | 10.325868 | 64 | 47.203968 |
| 15 | 11.063430 | 65 | 47.941530 |
| 16 | 11.800992 | 66 | 48.679092 |
| 17 | 12.538554 | 67 | 49.416654 |
| 18 | 13.276116 | 68 | 50.154216 |
| 19 | 14.013678 | 69 | 50.891778 |
| 20 | 14.751240 | 70 | 51.629340 |
| 21 | 15.488802 | 71 | 52.366902 |
| 22 | 16.226364 | 72 | 53.104464 |
| 23 | 16.963926 | 73 | 53.842026 |
| 24 | 17.701488 | 74 | 54.579588 |
| 25 | 18.439050 | 75 | 55.317150 |
| 26 | 19.176612 | 76 | 56.054712 |
| 27 | 19.914174 | 77 | 56.792274 |
| 28 | 20.651736 | 78 | 57.529836 |
| 29 | 21.389298 | 79 | 58.267398 |
| 30 | 22.126860 | 80 | 59.004960 |
| 31 | 22.864422 | 81 | 59.742522 |
| 32 | 23.601984 | 82 | 60.480084 |
| 33 | 24.339546 | 83 | 61.217646 |
| 34 | 25.077108 | 84 | 61.955208 |
| 35 | 25.814670 | 85 | 62.692770 |
| 36 | 26.552232 | 86 | 63.430332 |
| 37 | 27.289794 | 87 | 64.167894 |
| 38 | 28.027356 | 88 | 64.905456 |
| 39 | 28.764918 | 89 | 65.643018 |
| 40 | 29.502480 | 90 | 66.380580 |
| 41 | 30.240042 | 91 | 67.118142 |
| 42 | 30.977604 | 92 | 67.855704 |
| 43 | 31.715166 | 93 | 68.593266 |
| 44 | 32.452728 | 94 | 69.330828 |
| 45 | 33.190290 | 95 | 70.068390 |
| 46 | 33.927852 | 96 | 70.805952 |
| 47 | 34.665414 | 97 | 71.543514 |
| 48 | 35.402976 | 98 | 72.281076 |
| 49 | 36.140538 | 99 | 73.018638 |
| 50 | 36.878100 | 100 | 73.756200 |

## POWER
### Foot-pounds per Minute to Milliwatts

| ft·lb/min | mW | ft·lb/min | mW |
|---|---|---|---|
| 1 | 22.596970 | 51 | 1152.445470 |
| 2 | 45.193940 | 52 | 1175.042440 |
| 3 | 67.790910 | 53 | 1197.639410 |
| 4 | 90.387880 | 54 | 1220.236380 |
| 5 | 112.984850 | 55 | 1242.833350 |
| 6 | 135.581820 | 56 | 1265.430320 |
| 7 | 158.178790 | 57 | 1288.027290 |
| 8 | 180.775760 | 58 | 1310.624260 |
| 9 | 203.372730 | 59 | 1333.221230 |
| 10 | 225.969700 | 60 | 1355.818200 |
| 11 | 248.566670 | 61 | 1378.415170 |
| 12 | 271.163640 | 62 | 1401.012140 |
| 13 | 293.760610 | 63 | 1423.609110 |
| 14 | 316.357580 | 64 | 1446.206080 |
| 15 | 338.954550 | 65 | 1468.803050 |
| 16 | 361.551520 | 66 | 1491.400020 |
| 17 | 384.148490 | 67 | 1513.996990 |
| 18 | 406.745460 | 68 | 1536.593960 |
| 19 | 429.342430 | 69 | 1559.190930 |
| 20 | 451.939400 | 70 | 1581.787900 |
| 21 | 474.536370 | 71 | 1604.384870 |
| 22 | 497.133340 | 72 | 1626.981840 |
| 23 | 519.730310 | 73 | 1649.578810 |
| 24 | 542.327280 | 74 | 1672.175780 |
| 25 | 564.924250 | 75 | 1694.772750 |
| 26 | 587.521220 | 76 | 1717.369720 |
| 27 | 610.118190 | 77 | 1739.966690 |
| 28 | 632.715160 | 78 | 1762.563660 |
| 29 | 655.312130 | 79 | 1785.160630 |
| 30 | 677.909100 | 80 | 1807.757600 |
| 31 | 700.506070 | 81 | 1830.354570 |
| 32 | 723.103040 | 82 | 1852.951540 |
| 33 | 745.700010 | 83 | 1875.548510 |
| 34 | 768.296980 | 84 | 1898.145480 |
| 35 | 790.893950 | 85 | 1920.742450 |
| 36 | 813.490920 | 86 | 1943.339420 |
| 37 | 836.087890 | 87 | 1965.936390 |
| 38 | 858.684860 | 88 | 1988.533360 |
| 39 | 881.281830 | 89 | 2011.130330 |
| 40 | 903.878800 | 90 | 2033.727300 |
| 41 | 926.475770 | 91 | 2056.324270 |
| 42 | 949.072740 | 92 | 2078.921240 |
| 43 | 971.669710 | 93 | 2101.518210 |
| 44 | 994.266680 | 94 | 2124.115180 |
| 45 | 1016.863650 | 95 | 2146.712150 |
| 46 | 1039.460620 | 96 | 2169.309120 |
| 47 | 1062.057590 | 97 | 2191.906090 |
| 48 | 1084.654560 | 98 | 2214.503060 |
| 49 | 1107.251530 | 99 | 2237.100030 |
| 50 | 1129.848500 | 100 | 2259.697000 |

## POWER
### Milliwatts to Foot-pounds per Minute

| mW | ft·lb/min | mW | ft·lb/min |
|---|---|---|---|
| 1 | .044254 | 51 | 2.256954 |
| 2 | .088508 | 52 | 2.301208 |
| 3 | .132762 | 53 | 2.345462 |
| 4 | .177016 | 54 | 2.389716 |
| 5 | .221270 | 55 | 2.433970 |
| 6 | .265524 | 56 | 2.478224 |
| 7 | .309778 | 57 | 2.522478 |
| 8 | .354032 | 58 | 2.566732 |
| 9 | .398286 | 59 | 2.610986 |
| 10 | .442540 | 60 | 2.655240 |
| 11 | .486794 | 61 | 2.699494 |
| 12 | .531048 | 62 | 2.743748 |
| 13 | .575302 | 63 | 2.788002 |
| 14 | .619556 | 64 | 2.832256 |
| 15 | .663810 | 65 | 2.876510 |
| 16 | .708064 | 66 | 2.920764 |
| 17 | .752318 | 67 | 2.965018 |
| 18 | .796572 | 68 | 3.009272 |
| 19 | .840826 | 69 | 3.053526 |
| 20 | .885080 | 70 | 3.097780 |
| 21 | .929334 | 71 | 3.142034 |
| 22 | .973588 | 72 | 3.186288 |
| 23 | 1.017842 | 73 | 3.230542 |
| 24 | 1.062096 | 74 | 3.274796 |
| 25 | 1.106350 | 75 | 3.319050 |
| 26 | 1.150604 | 76 | 3.363304 |
| 27 | 1.194858 | 77 | 3.407558 |
| 28 | 1.239112 | 78 | 3.451812 |
| 29 | 1.283366 | 79 | 3.496066 |
| 30 | 1.327620 | 80 | 3.540320 |
| 31 | 1.371874 | 81 | 3.584574 |
| 32 | 1.416128 | 82 | 3.628828 |
| 33 | 1.460382 | 83 | 3.673082 |
| 34 | 1.504636 | 84 | 3.717336 |
| 35 | 1.548890 | 85 | 3.761590 |
| 36 | 1.593144 | 86 | 3.805844 |
| 37 | 1.637398 | 87 | 3.850098 |
| 38 | 1.681652 | 88 | 3.894352 |
| 39 | 1.725906 | 89 | 3.938606 |
| 40 | 1.770160 | 90 | 3.982860 |
| 41 | 1.814414 | 91 | 4.027114 |
| 42 | 1.858668 | 92 | 4.071368 |
| 43 | 1.902922 | 93 | 4.115622 |
| 44 | 1.947176 | 94 | 4.159876 |
| 45 | 1.991430 | 95 | 4.204130 |
| 46 | 2.035684 | 96 | 4.248384 |
| 47 | 2.079938 | 97 | 4.292638 |
| 48 | 2.124192 | 98 | 4.336892 |
| 49 | 2.168446 | 99 | 4.381146 |
| 50 | 2.212700 | 100 | 4.425400 |

## POWER
### Foot-pounds per Hour to Milliwatts

| ft·lb/h | mW | ft·lb/h | mW |
|---|---|---|---|
| 1 | .376616 | 51 | 19.207416 |
| 2 | .753232 | 52 | 19.584032 |
| 3 | 1.129848 | 53 | 19.960648 |
| 4 | 1.506464 | 54 | 20.337264 |
| 5 | 1.883080 | 55 | 20.713880 |
| 6 | 2.259696 | 56 | 21.090496 |
| 7 | 2.636312 | 57 | 21.467112 |
| 8 | 3.012928 | 58 | 21.843728 |
| 9 | 3.389544 | 59 | 22.220344 |
| 10 | 3.766160 | 60 | 22.596960 |
| 11 | 4.142776 | 61 | 22.973576 |
| 12 | 4.519392 | 62 | 23.350192 |
| 13 | 4.896008 | 63 | 23.726808 |
| 14 | 5.272624 | 64 | 24.103424 |
| 15 | 5.649240 | 65 | 24.480040 |
| 16 | 6.025856 | 66 | 24.856656 |
| 17 | 6.402472 | 67 | 25.233272 |
| 18 | 6.779088 | 68 | 25.609888 |
| 19 | 7.155704 | 69 | 25.986504 |
| 20 | 7.532320 | 70 | 26.363120 |
| 21 | 7.908936 | 71 | 26.739736 |
| 22 | 8.285552 | 72 | 27.116352 |
| 23 | 8.662168 | 73 | 27.492968 |
| 24 | 9.038784 | 74 | 27.869584 |
| 25 | 9.415400 | 75 | 28.246200 |
| 26 | 9.792016 | 76 | 28.622816 |
| 27 | 10.168632 | 77 | 28.999432 |
| 28 | 10.545248 | 78 | 29.376048 |
| 29 | 10.921864 | 79 | 29.752664 |
| 30 | 11.298480 | 80 | 30.129280 |
| 31 | 11.675096 | 81 | 30.505896 |
| 32 | 12.051712 | 82 | 30.882512 |
| 33 | 12.428328 | 83 | 31.259128 |
| 34 | 12.804944 | 84 | 31.635744 |
| 35 | 13.181560 | 85 | 32.012360 |
| 36 | 13.558176 | 86 | 32.388976 |
| 37 | 13.934792 | 87 | 32.765592 |
| 38 | 14.311408 | 88 | 33.142208 |
| 39 | 14.688024 | 89 | 33.518824 |
| 40 | 15.064640 | 90 | 33.895440 |
| 41 | 15.441256 | 91 | 34.272056 |
| 42 | 15.817872 | 92 | 34.648672 |
| 43 | 16.194488 | 93 | 35.025288 |
| 44 | 16.571104 | 94 | 35.401904 |
| 45 | 16.947720 | 95 | 35.778520 |
| 46 | 17.324336 | 96 | 36.155136 |
| 47 | 17.700952 | 97 | 36.531752 |
| 48 | 18.077568 | 98 | 36.908368 |
| 49 | 18.454184 | 99 | 37.284984 |
| 50 | 18.830800 | 100 | 37.661600 |

## POWER
### Milliwatts to Foot-pounds per Hour

| mW | ft·lb/h | mW | ft·lb/h |
|---|---|---|---|
| 1 | 2.655224 | 51 | 135.416424 |
| 2 | 5.310448 | 52 | 138.071648 |
| 3 | 7.965672 | 53 | 140.726872 |
| 4 | 10.620896 | 54 | 143.382096 |
| 5 | 13.276120 | 55 | 146.037320 |
| 6 | 15.931344 | 56 | 148.692544 |
| 7 | 18.586568 | 57 | 151.347768 |
| 8 | 21.241792 | 58 | 154.002992 |
| 9 | 23.897016 | 59 | 156.658216 |
| 10 | 26.552240 | 60 | 159.313440 |
| 11 | 29.207464 | 61 | 161.968664 |
| 12 | 31.862688 | 62 | 164.623888 |
| 13 | 34.517912 | 63 | 167.279112 |
| 14 | 37.173136 | 64 | 169.934336 |
| 15 | 39.828360 | 65 | 172.589560 |
| 16 | 42.483584 | 66 | 175.244784 |
| 17 | 45.138808 | 67 | 177.900008 |
| 18 | 47.794032 | 68 | 180.555232 |
| 19 | 50.449256 | 69 | 183.210456 |
| 20 | 53.104480 | 70 | 185.865680 |
| 21 | 55.759704 | 71 | 188.520904 |
| 22 | 58.414928 | 72 | 191.176128 |
| 23 | 61.070152 | 73 | 193.831352 |
| 24 | 63.725376 | 74 | 196.486576 |
| 25 | 66.380600 | 75 | 199.141800 |
| 26 | 69.035824 | 76 | 201.797024 |
| 27 | 71.691048 | 77 | 204.452248 |
| 28 | 74.346272 | 78 | 207.107472 |
| 29 | 77.001496 | 79 | 209.762696 |
| 30 | 79.656720 | 80 | 212.417920 |
| 31 | 82.311944 | 81 | 215.073144 |
| 32 | 84.967168 | 82 | 217.728368 |
| 33 | 87.622392 | 83 | 220.383592 |
| 34 | 90.277616 | 84 | 223.038816 |
| 35 | 92.932840 | 85 | 225.694040 |
| 36 | 95.588064 | 86 | 228.349264 |
| 37 | 98.243288 | 87 | 231.004488 |
| 38 | 100.898512 | 88 | 233.659712 |
| 39 | 103.553736 | 89 | 236.314936 |
| 40 | 106.208960 | 90 | 238.970160 |
| 41 | 108.864184 | 91 | 241.625384 |
| 42 | 111.519408 | 92 | 244.280608 |
| 43 | 114.174632 | 93 | 246.935832 |
| 44 | 116.829856 | 94 | 249.591056 |
| 45 | 119.485080 | 95 | 252.246280 |
| 46 | 122.140304 | 96 | 254.901504 |
| 47 | 124.795528 | 97 | 257.556728 |
| 48 | 127.450752 | 98 | 260.211952 |
| 49 | 130.105976 | 99 | 262.867176 |
| 50 | 132.761200 | 100 | 265.522400 |

## POWER
### Meter-kilograms per Second to Watts

| m·kg/s | W | m·kg/s | W |
|---|---|---|---|
| 1 | 9.806650 | 51 | 500.139150 |
| 2 | 19.613300 | 52 | 509.945800 |
| 3 | 29.419950 | 53 | 519.752450 |
| 4 | 39.226600 | 54 | 529.559100 |
| 5 | 49.033250 | 55 | 539.365750 |
| 6 | 58.839900 | 56 | 549.172400 |
| 7 | 68.646550 | 57 | 558.979050 |
| 8 | 78.453200 | 58 | 568.785700 |
| 9 | 88.259850 | 59 | 578.592350 |
| 10 | 98.066500 | 60 | 588.399000 |
| 11 | 107.873150 | 61 | 598.205650 |
| 12 | 117.679800 | 62 | 608.012300 |
| 13 | 127.486450 | 63 | 617.818950 |
| 14 | 137.293100 | 64 | 627.625600 |
| 15 | 147.099750 | 65 | 637.432250 |
| 16 | 156.906400 | 66 | 647.238900 |
| 17 | 166.713050 | 67 | 657.045550 |
| 18 | 176.519700 | 68 | 666.852200 |
| 19 | 186.326350 | 69 | 676.658850 |
| 20 | 196.133000 | 70 | 686.465500 |
| 21 | 205.939650 | 71 | 696.272150 |
| 22 | 215.746300 | 72 | 706.078800 |
| 23 | 225.552950 | 73 | 715.885450 |
| 24 | 235.359600 | 74 | 725.692100 |
| 25 | 245.166250 | 75 | 735.498750 |
| 26 | 254.972900 | 76 | 745.305400 |
| 27 | 264.779550 | 77 | 755.112050 |
| 28 | 274.586200 | 78 | 764.918700 |
| 29 | 284.392850 | 79 | 774.725350 |
| 30 | 294.199500 | 80 | 784.532000 |
| 31 | 304.006150 | 81 | 794.338650 |
| 32 | 313.812800 | 82 | 804.145300 |
| 33 | 323.619450 | 83 | 813.951950 |
| 34 | 333.426100 | 84 | 823.758600 |
| 35 | 343.232750 | 85 | 833.565250 |
| 36 | 353.039400 | 86 | 843.371900 |
| 37 | 362.846050 | 87 | 853.178550 |
| 38 | 372.652700 | 88 | 862.985200 |
| 39 | 382.459350 | 89 | 872.791850 |
| 40 | 392.266000 | 90 | 882.598500 |
| 41 | 402.072650 | 91 | 892.405150 |
| 42 | 411.879300 | 92 | 902.211800 |
| 43 | 421.685950 | 93 | 912.018450 |
| 44 | 431.492600 | 94 | 921.825100 |
| 45 | 441.299250 | 95 | 931.631750 |
| 46 | 451.105900 | 96 | 941.438400 |
| 47 | 460.912550 | 97 | 951.245050 |
| 48 | 470.719200 | 98 | 961.051700 |
| 49 | 480.525850 | 99 | 970.858350 |
| 50 | 490.332500 | 100 | 980.665000 |

## POWER
### Watts to Meter-kilograms per Second

| W | m·kg/s | W | m·kg/s |
|---|---|---|---|
| 1 | .101972 | 51 | 5.200572 |
| 2 | .203944 | 52 | 5.302544 |
| 3 | .305916 | 53 | 5.404516 |
| 4 | .407888 | 54 | 5.506488 |
| 5 | .509860 | 55 | 5.608460 |
| 6 | .611832 | 56 | 5.710432 |
| 7 | .713804 | 57 | 5.812404 |
| 8 | .815776 | 58 | 5.914376 |
| 9 | .917748 | 59 | 6.016348 |
| 10 | 1.019720 | 60 | 6.118320 |
| 11 | 1.121692 | 61 | 6.220292 |
| 12 | 1.223664 | 62 | 6.322264 |
| 13 | 1.325636 | 63 | 6.424236 |
| 14 | 1.427608 | 64 | 6.526208 |
| 15 | 1.529580 | 65 | 6.628180 |
| 16 | 1.631552 | 66 | 6.730152 |
| 17 | 1.733524 | 67 | 6.832124 |
| 18 | 1.835496 | 68 | 6.934096 |
| 19 | 1.937468 | 69 | 7.036068 |
| 20 | 2.039440 | 70 | 7.138040 |
| 21 | 2.141412 | 71 | 7.240012 |
| 22 | 2.243384 | 72 | 7.341984 |
| 23 | 2.345356 | 73 | 7.443956 |
| 24 | 2.447328 | 74 | 7.545928 |
| 25 | 2.549300 | 75 | 7.647900 |
| 26 | 2.651272 | 76 | 7.749872 |
| 27 | 2.753244 | 77 | 7.851844 |
| 28 | 2.855216 | 78 | 7.953816 |
| 29 | 2.957188 | 79 | 8.055788 |
| 30 | 3.059160 | 80 | 8.157760 |
| 31 | 3.161132 | 81 | 8.259732 |
| 32 | 3.263104 | 82 | 8.361704 |
| 33 | 3.365076 | 83 | 8.463676 |
| 34 | 3.467048 | 84 | 8.565648 |
| 35 | 3.569020 | 85 | 8.667620 |
| 36 | 3.670992 | 86 | 8.769592 |
| 37 | 3.772964 | 87 | 8.871564 |
| 38 | 3.874936 | 88 | 8.973536 |
| 39 | 3.976908 | 89 | 9.075508 |
| 40 | 4.078880 | 90 | 9.177480 |
| 41 | 4.180852 | 91 | 9.279452 |
| 42 | 4.282824 | 92 | 9.381424 |
| 43 | 4.384796 | 93 | 9.483396 |
| 44 | 4.486768 | 94 | 9.585368 |
| 45 | 4.588740 | 95 | 9.687340 |
| 46 | 4.690712 | 96 | 9.789312 |
| 47 | 4.792684 | 97 | 9.891284 |
| 48 | 4.894656 | 98 | 9.993256 |
| 49 | 4.996628 | 99 | 10.095228 |
| 50 | 5.098600 | 100 | 10.197200 |

## POWER
## British Thermal Units per Second to Kilowatts

| Btu/s | kW | Btu/s | kW |
|---|---|---|---|
| 1 | 1.054800 | 51 | 53.794800 |
| 2 | 2.109600 | 52 | 54.849600 |
| 3 | 3.164400 | 53 | 55.904400 |
| 4 | 4.219200 | 54 | 56.959200 |
| 5 | 5.274000 | 55 | 58.014000 |
| 6 | 6.328800 | 56 | 59.068800 |
| 7 | 7.383600 | 57 | 60.123600 |
| 8 | 8.438400 | 58 | 61.178400 |
| 9 | 9.493200 | 59 | 62.233200 |
| 10 | 10.548000 | 60 | 63.288000 |
| 11 | 11.602800 | 61 | 64.342800 |
| 12 | 12.657600 | 62 | 65.397600 |
| 13 | 13.712400 | 63 | 66.452400 |
| 14 | 14.767200 | 64 | 67.507200 |
| 15 | 15.822000 | 65 | 68.562000 |
| 16 | 16.876800 | 66 | 69.616800 |
| 17 | 17.931600 | 67 | 70.671600 |
| 18 | 18.986400 | 68 | 71.726400 |
| 19 | 20.041200 | 69 | 72.781200 |
| 20 | 21.096000 | 70 | 73.836000 |
| 21 | 22.150800 | 71 | 74.890800 |
| 22 | 23.205600 | 72 | 75.945600 |
| 23 | 24.260400 | 73 | 77.000400 |
| 24 | 25.315200 | 74 | 78.055200 |
| 25 | 26.370000 | 75 | 79.110000 |
| 26 | 27.424800 | 76 | 80.164800 |
| 27 | 28.479600 | 77 | 81.219600 |
| 28 | 29.534400 | 78 | 82.274400 |
| 29 | 30.589200 | 79 | 83.329200 |
| 30 | 31.644000 | 80 | 84.384000 |
| 31 | 32.698800 | 81 | 85.438800 |
| 32 | 33.753600 | 82 | 86.493600 |
| 33 | 34.808400 | 83 | 87.548400 |
| 34 | 35.863200 | 84 | 88.603200 |
| 35 | 36.918000 | 85 | 89.658000 |
| 36 | 37.972800 | 86 | 90.712800 |
| 37 | 39.027600 | 87 | 91.767600 |
| 38 | 40.082400 | 88 | 92.822400 |
| 39 | 41.137200 | 89 | 93.877200 |
| 40 | 42.192000 | 90 | 94.932000 |
| 41 | 43.246800 | 91 | 95.986800 |
| 42 | 44.301600 | 92 | 97.041600 |
| 43 | 45.356400 | 93 | 98.096400 |
| 44 | 46.411200 | 94 | 99.151200 |
| 45 | 47.466000 | 95 | 100.206000 |
| 46 | 48.520800 | 96 | 101.260800 |
| 47 | 49.575600 | 97 | 102.315600 |
| 48 | 50.630400 | 98 | 103.370400 |
| 49 | 51.685200 | 99 | 104.425200 |
| 50 | 52.740000 | 100 | 105.480000 |

## POWER
## Kilowatts to British Thermal Units per Second

| kW | Btu/s | kW | Btu/s |
|---|---|---|---|
| 1 | .948047 | 51 | 48.350397 |
| 2 | 1.896094 | 52 | 49.298444 |
| 3 | 2.844141 | 53 | 50.246491 |
| 4 | 3.792188 | 54 | 51.194538 |
| 5 | 4.740235 | 55 | 52.142585 |
| 6 | 5.688282 | 56 | 53.090632 |
| 7 | 6.636329 | 57 | 54.038679 |
| 8 | 7.584376 | 58 | 54.986726 |
| 9 | 8.532423 | 59 | 55.934773 |
| 10 | 9.480470 | 60 | 56.882820 |
| 11 | 10.428517 | 61 | 57.830867 |
| 12 | 11.376564 | 62 | 58.778914 |
| 13 | 12.324611 | 63 | 59.726961 |
| 14 | 13.272658 | 64 | 60.675008 |
| 15 | 14.220705 | 65 | 61.623055 |
| 16 | 15.168752 | 66 | 62.571102 |
| 17 | 16.116799 | 67 | 63.519149 |
| 18 | 17.064846 | 68 | 64.467196 |
| 19 | 18.012893 | 69 | 65.415243 |
| 20 | 18.960940 | 70 | 66.363290 |
| 21 | 19.908987 | 71 | 67.311337 |
| 22 | 20.857034 | 72 | 68.259384 |
| 23 | 21.805081 | 73 | 69.207431 |
| 24 | 22.753128 | 74 | 70.155478 |
| 25 | 23.701175 | 75 | 71.103525 |
| 26 | 24.649222 | 76 | 72.051572 |
| 27 | 25.597269 | 77 | 72.999619 |
| 28 | 26.545316 | 78 | 73.947666 |
| 29 | 27.493363 | 79 | 74.895713 |
| 30 | 28.441410 | 80 | 75.843760 |
| 31 | 29.389457 | 81 | 76.791807 |
| 32 | 30.337504 | 82 | 77.739854 |
| 33 | 31.285551 | 83 | 78.687901 |
| 34 | 32.233598 | 84 | 79.635948 |
| 35 | 33.181645 | 85 | 80.583995 |
| 36 | 34.129692 | 86 | 81.532042 |
| 37 | 35.077739 | 87 | 82.480089 |
| 38 | 36.025786 | 88 | 83.428136 |
| 39 | 36.973833 | 89 | 84.376183 |
| 40 | 37.921880 | 90 | 85.324230 |
| 41 | 38.869927 | 91 | 86.272277 |
| 42 | 39.817974 | 92 | 87.220324 |
| 43 | 40.766021 | 93 | 88.168371 |
| 44 | 41.714068 | 94 | 89.116418 |
| 45 | 42.662115 | 95 | 90.064465 |
| 46 | 43.610162 | 96 | 91.012512 |
| 47 | 44.558209 | 97 | 91.960559 |
| 48 | 45.506256 | 98 | 92.908606 |
| 49 | 46.454303 | 99 | 93.856653 |
| 50 | 47.402350 | 100 | 94.804700 |

## POWER
## British Thermal Units per Hour
## to
## Watts

| Btu/h | W | Btu/h | W |
|---|---|---|---|
| 1 | .293000 | 51 | 14.943000 |
| 2 | .586000 | 52 | 15.236000 |
| 3 | .879000 | 53 | 15.529000 |
| 4 | 1.172000 | 54 | 15.822000 |
| 5 | 1.465000 | 55 | 16.115000 |
| 6 | 1.758000 | 56 | 16.408000 |
| 7 | 2.051000 | 57 | 16.701000 |
| 8 | 2.344000 | 58 | 16.994000 |
| 9 | 2.637000 | 59 | 17.287000 |
| 10 | 2.930000 | 60 | 17.580000 |
| 11 | 3.223000 | 61 | 17.873000 |
| 12 | 3.516000 | 62 | 18.166000 |
| 13 | 3.809000 | 63 | 18.459000 |
| 14 | 4.102000 | 64 | 18.752000 |
| 15 | 4.395000 | 65 | 19.045000 |
| 16 | 4.688000 | 66 | 19.338000 |
| 17 | 4.981000 | 67 | 19.631000 |
| 18 | 5.274000 | 68 | 19.924000 |
| 19 | 5.567000 | 69 | 20.217000 |
| 20 | 5.860000 | 70 | 20.510000 |
| 21 | 6.153000 | 71 | 20.803000 |
| 22 | 6.446000 | 72 | 21.096000 |
| 23 | 6.739000 | 73 | 21.389000 |
| 24 | 7.032000 | 74 | 21.682000 |
| 25 | 7.325000 | 75 | 21.975000 |
| 26 | 7.618000 | 76 | 22.268000 |
| 27 | 7.911000 | 77 | 22.561000 |
| 28 | 8.204000 | 78 | 22.854000 |
| 29 | 8.497000 | 79 | 23.147000 |
| 30 | 8.790000 | 80 | 23.440000 |
| 31 | 9.083000 | 81 | 23.733000 |
| 32 | 9.376000 | 82 | 24.026000 |
| 33 | 9.669000 | 83 | 24.319000 |
| 34 | 9.962000 | 84 | 24.612000 |
| 35 | 10.255000 | 85 | 24.905000 |
| 36 | 10.548000 | 86 | 25.198000 |
| 37 | 10.841000 | 87 | 25.491000 |
| 38 | 11.134000 | 88 | 25.784000 |
| 39 | 11.427000 | 89 | 26.077000 |
| 40 | 11.720000 | 90 | 26.370000 |
| 41 | 12.013000 | 91 | 26.663000 |
| 42 | 12.306000 | 92 | 26.956000 |
| 43 | 12.599000 | 93 | 27.249000 |
| 44 | 12.892000 | 94 | 27.542000 |
| 45 | 13.185000 | 95 | 27.835000 |
| 46 | 13.478000 | 96 | 28.128000 |
| 47 | 13.771000 | 97 | 28.421000 |
| 48 | 14.064000 | 98 | 28.714000 |
| 49 | 14.357000 | 99 | 29.007000 |
| 50 | 14.650000 | 100 | 29.300000 |

## POWER
## Watts
## to
## British Thermal Units per Hour

| W | Btu/h | W | Btu/h |
|---|---|---|---|
| 1 | 3.412969 | 51 | 174.061419 |
| 2 | 6.825938 | 52 | 177.474388 |
| 3 | 10.238907 | 53 | 180.887357 |
| 4 | 13.651876 | 54 | 184.300326 |
| 5 | 17.064845 | 55 | 187.713295 |
| 6 | 20.477814 | 56 | 191.126264 |
| 7 | 23.890783 | 57 | 194.539233 |
| 8 | 27.303752 | 58 | 197.952202 |
| 9 | 30.716721 | 59 | 201.365171 |
| 10 | 34.129690 | 60 | 204.778140 |
| 11 | 37.542659 | 61 | 208.191109 |
| 12 | 40.955628 | 62 | 211.604078 |
| 13 | 44.368597 | 63 | 215.017047 |
| 14 | 47.781566 | 64 | 218.430016 |
| 15 | 51.194535 | 65 | 221.842985 |
| 16 | 54.607504 | 66 | 225.255954 |
| 17 | 58.020473 | 67 | 228.668923 |
| 18 | 61.433442 | 68 | 232.081892 |
| 19 | 64.846411 | 69 | 235.494861 |
| 20 | 68.259380 | 70 | 238.907830 |
| 21 | 71.672349 | 71 | 242.320799 |
| 22 | 75.085318 | 72 | 245.733768 |
| 23 | 78.498287 | 73 | 249.146737 |
| 24 | 81.911256 | 74 | 252.559706 |
| 25 | 85.324225 | 75 | 255.972675 |
| 26 | 88.737194 | 76 | 259.385644 |
| 27 | 92.150163 | 77 | 262.798613 |
| 28 | 95.563132 | 78 | 266.211582 |
| 29 | 98.976101 | 79 | 269.624551 |
| 30 | 102.389070 | 80 | 273.037520 |
| 31 | 105.802039 | 81 | 276.450489 |
| 32 | 109.215008 | 82 | 279.863458 |
| 33 | 112.627977 | 83 | 283.276427 |
| 34 | 116.040946 | 84 | 286.689396 |
| 35 | 119.453915 | 85 | 290.102365 |
| 36 | 122.866884 | 86 | 293.515334 |
| 37 | 126.279853 | 87 | 296.928303 |
| 38 | 129.692822 | 88 | 300.341272 |
| 39 | 133.105791 | 89 | 303.754241 |
| 40 | 136.518760 | 90 | 307.167210 |
| 41 | 139.931729 | 91 | 310.580179 |
| 42 | 143.344698 | 92 | 313.993148 |
| 43 | 146.757667 | 93 | 317.406117 |
| 44 | 150.170636 | 94 | 320.819086 |
| 45 | 153.583605 | 95 | 324.232055 |
| 46 | 156.996574 | 96 | 327.645024 |
| 47 | 160.409543 | 97 | 331.057993 |
| 48 | 163.822512 | 98 | 334.470962 |
| 49 | 167.235481 | 99 | 337.883931 |
| 50 | 170.648450 | 100 | 341.296900 |

## POWER
### Calories per Hour to Milliwatts

| cal/h | mW | cal/h | mW |
|---|---|---|---|
| 1 | 1.162917 | 51 | 59.308767 |
| 2 | 2.325834 | 52 | 60.471684 |
| 3 | 3.488751 | 53 | 61.634601 |
| 4 | 4.651668 | 54 | 62.797518 |
| 5 | 5.814585 | 55 | 63.960435 |
| 6 | 6.977502 | 56 | 65.123352 |
| 7 | 8.140419 | 57 | 66.286269 |
| 8 | 9.303336 | 58 | 67.449186 |
| 9 | 10.466253 | 59 | 68.612103 |
| 10 | 11.629170 | 60 | 69.775020 |
| 11 | 12.792087 | 61 | 70.937937 |
| 12 | 13.955004 | 62 | 72.100854 |
| 13 | 15.117921 | 63 | 73.263771 |
| 14 | 16.280838 | 64 | 74.426688 |
| 15 | 17.443755 | 65 | 75.589605 |
| 16 | 18.606672 | 66 | 76.752522 |
| 17 | 19.769589 | 67 | 77.915439 |
| 18 | 20.932506 | 68 | 79.078356 |
| 19 | 22.095423 | 69 | 80.241273 |
| 20 | 23.258340 | 70 | 81.404190 |
| 21 | 24.421257 | 71 | 82.567107 |
| 22 | 25.584174 | 72 | 83.730024 |
| 23 | 26.747091 | 73 | 84.892941 |
| 24 | 27.910008 | 74 | 86.055858 |
| 25 | 29.072925 | 75 | 87.218775 |
| 26 | 30.235842 | 76 | 88.381692 |
| 27 | 31.398759 | 77 | 89.544609 |
| 28 | 32.561676 | 78 | 90.707526 |
| 29 | 33.724593 | 79 | 91.870443 |
| 30 | 34.887510 | 80 | 93.033360 |
| 31 | 36.050427 | 81 | 94.196277 |
| 32 | 37.213344 | 82 | 95.359194 |
| 33 | 38.376261 | 83 | 96.522111 |
| 34 | 39.539178 | 84 | 97.685028 |
| 35 | 40.702095 | 85 | 98.847945 |
| 36 | 41.865012 | 86 | 100.010862 |
| 37 | 43.027929 | 87 | 101.173779 |
| 38 | 44.190846 | 88 | 102.336696 |
| 39 | 45.353763 | 89 | 103.499613 |
| 40 | 46.516680 | 90 | 104.662530 |
| 41 | 47.679597 | 91 | 105.825447 |
| 42 | 48.842514 | 92 | 106.988364 |
| 43 | 50.005431 | 93 | 108.151281 |
| 44 | 51.168348 | 94 | 109.314198 |
| 45 | 52.331265 | 95 | 110.477115 |
| 46 | 53.494182 | 96 | 111.640032 |
| 47 | 54.657099 | 97 | 112.802949 |
| 48 | 55.820016 | 98 | 113.965866 |
| 49 | 56.982933 | 99 | 115.128783 |
| 50 | 58.145850 | 100 | 116.291700 |

## POWER
### Milliwatts to Calories per Hour

| mW | cal/h | mW | cal/h |
|---|---|---|---|
| 1 | .859907 | 51 | 43.855257 |
| 2 | 1.719814 | 52 | 44.715164 |
| 3 | 2.579721 | 53 | 45.575071 |
| 4 | 3.439628 | 54 | 46.434978 |
| 5 | 4.299535 | 55 | 47.294885 |
| 6 | 5.159442 | 56 | 48.154792 |
| 7 | 6.019349 | 57 | 49.014699 |
| 8 | 6.879256 | 58 | 49.874606 |
| 9 | 7.739163 | 59 | 50.734513 |
| 10 | 8.599070 | 60 | 51.594420 |
| 11 | 9.458977 | 61 | 52.454327 |
| 12 | 10.318884 | 62 | 53.314234 |
| 13 | 11.178791 | 63 | 54.174141 |
| 14 | 12.038698 | 64 | 55.034048 |
| 15 | 12.898605 | 65 | 55.893955 |
| 16 | 13.758512 | 66 | 56.753862 |
| 17 | 14.618419 | 67 | 57.613769 |
| 18 | 15.478326 | 68 | 58.473676 |
| 19 | 16.338233 | 69 | 59.333583 |
| 20 | 17.198140 | 70 | 60.193490 |
| 21 | 18.058047 | 71 | 61.053397 |
| 22 | 18.917954 | 72 | 61.913304 |
| 23 | 19.777861 | 73 | 62.773211 |
| 24 | 20.637768 | 74 | 63.633118 |
| 25 | 21.497675 | 75 | 64.493025 |
| 26 | 22.357582 | 76 | 65.352932 |
| 27 | 23.217489 | 77 | 66.212839 |
| 28 | 24.077396 | 78 | 67.072746 |
| 29 | 24.937303 | 79 | 67.932653 |
| 30 | 25.797210 | 80 | 68.792560 |
| 31 | 26.657117 | 81 | 69.652467 |
| 32 | 27.517024 | 82 | 70.512374 |
| 33 | 28.376931 | 83 | 71.372281 |
| 34 | 29.236838 | 84 | 72.232188 |
| 35 | 30.096745 | 85 | 73.092095 |
| 36 | 30.956652 | 86 | 73.952002 |
| 37 | 31.816559 | 87 | 74.811909 |
| 38 | 32.676466 | 88 | 75.671816 |
| 39 | 33.536373 | 89 | 76.531723 |
| 40 | 34.396280 | 90 | 77.391630 |
| 41 | 35.256187 | 91 | 78.251537 |
| 42 | 36.116094 | 92 | 79.111444 |
| 43 | 36.976001 | 93 | 79.971351 |
| 44 | 37.835908 | 94 | 80.831258 |
| 45 | 38.695815 | 95 | 81.691165 |
| 46 | 39.555722 | 96 | 82.551072 |
| 47 | 40.415629 | 97 | 83.410979 |
| 48 | 41.275536 | 98 | 84.270886 |
| 49 | 42.135443 | 99 | 85.130793 |
| 50 | 42.995350 | 100 | 85.990700 |

## POWER
## Kilocalories per Second to Kilowatts

| kcal/s | kW | kcal/s | kW |
|---|---|---|---|
| 1 | 4.186000 | 51 | 213.486000 |
| 2 | 8.372000 | 52 | 217.672000 |
| 3 | 12.558000 | 53 | 221.858000 |
| 4 | 16.744000 | 54 | 226.044000 |
| 5 | 20.930000 | 55 | 230.230000 |
| 6 | 25.116000 | 56 | 234.416000 |
| 7 | 29.302000 | 57 | 238.602000 |
| 8 | 33.488000 | 58 | 242.788000 |
| 9 | 37.674000 | 59 | 246.974000 |
| 10 | 41.860000 | 60 | 251.160000 |
| 11 | 46.046000 | 61 | 255.346000 |
| 12 | 50.232000 | 62 | 259.532000 |
| 13 | 54.418000 | 63 | 263.718000 |
| 14 | 58.604000 | 64 | 267.904000 |
| 15 | 62.790000 | 65 | 272.090000 |
| 16 | 66.976000 | 66 | 276.276000 |
| 17 | 71.162000 | 67 | 280.462000 |
| 18 | 75.348000 | 68 | 284.648000 |
| 19 | 79.534000 | 69 | 288.834000 |
| 20 | 83.720000 | 70 | 293.020000 |
| 21 | 87.906000 | 71 | 297.206000 |
| 22 | 92.092000 | 72 | 301.392000 |
| 23 | 96.278000 | 73 | 305.578000 |
| 24 | 100.464000 | 74 | 309.764000 |
| 25 | 104.650000 | 75 | 313.950000 |
| 26 | 108.836000 | 76 | 318.136000 |
| 27 | 113.022000 | 77 | 322.322000 |
| 28 | 117.208000 | 78 | 326.508000 |
| 29 | 121.394000 | 79 | 330.694000 |
| 30 | 125.580000 | 80 | 334.880000 |
| 31 | 129.766000 | 81 | 339.066000 |
| 32 | 133.952000 | 82 | 343.252000 |
| 33 | 138.138000 | 83 | 347.438000 |
| 34 | 142.324000 | 84 | 351.624000 |
| 35 | 146.510000 | 85 | 355.810000 |
| 36 | 150.696000 | 86 | 359.996000 |
| 37 | 154.882000 | 87 | 364.182000 |
| 38 | 159.068000 | 88 | 368.368000 |
| 39 | 163.254000 | 89 | 372.554000 |
| 40 | 167.440000 | 90 | 376.740000 |
| 41 | 171.626000 | 91 | 380.926000 |
| 42 | 175.812000 | 92 | 385.112000 |
| 43 | 179.998000 | 93 | 389.298000 |
| 44 | 184.184000 | 94 | 393.484000 |
| 45 | 188.370000 | 95 | 397.670000 |
| 46 | 192.556000 | 96 | 401.856000 |
| 47 | 196.742000 | 97 | 406.042000 |
| 48 | 200.928000 | 98 | 410.228000 |
| 49 | 205.114000 | 99 | 414.414000 |
| 50 | 209.300000 | 100 | 418.600000 |

## POWER
## Kilowatts
## to
## Kilocalories per Second

| kW | kcal/s | kW | kcal/s |
|---|---|---|---|
| 1 | .238892 | 51 | 12.183492 |
| 2 | .477784 | 52 | 12.422384 |
| 3 | .716676 | 53 | 12.661276 |
| 4 | .955568 | 54 | 12.900168 |
| 5 | 1.194460 | 55 | 13.139060 |
| 6 | 1.433352 | 56 | 13.377952 |
| 7 | 1.672244 | 57 | 13.616844 |
| 8 | 1.911136 | 58 | 13.855736 |
| 9 | 2.150028 | 59 | 14.094628 |
| 10 | 2.388920 | 60 | 14.333520 |
| 11 | 2.627812 | 61 | 14.572412 |
| 12 | 2.866704 | 62 | 14.811304 |
| 13 | 3.105596 | 63 | 15.050196 |
| 14 | 3.344488 | 64 | 15.289088 |
| 15 | 3.583380 | 65 | 15.527980 |
| 16 | 3.822272 | 66 | 15.766872 |
| 17 | 4.061164 | 67 | 16.005764 |
| 18 | 4.300056 | 68 | 16.244656 |
| 19 | 4.538948 | 69 | 16.483548 |
| 20 | 4.777840 | 70 | 16.722440 |
| 21 | 5.016732 | 71 | 16.961332 |
| 22 | 5.255624 | 72 | 17.200224 |
| 23 | 5.494516 | 73 | 17.439116 |
| 24 | 5.733408 | 74 | 17.678008 |
| 25 | 5.972300 | 75 | 17.916900 |
| 26 | 6.211192 | 76 | 18.155792 |
| 27 | 6.450084 | 77 | 18.394684 |
| 28 | 6.688976 | 78 | 18.633576 |
| 29 | 6.927868 | 79 | 18.872468 |
| 30 | 7.166760 | 80 | 19.111360 |
| 31 | 7.405652 | 81 | 19.350252 |
| 32 | 7.644544 | 82 | 19.589144 |
| 33 | 7.883436 | 83 | 19.828036 |
| 34 | 8.122328 | 84 | 20.066928 |
| 35 | 8.361220 | 85 | 20.305820 |
| 36 | 8.600112 | 86 | 20.544712 |
| 37 | 8.839004 | 87 | 20.783604 |
| 38 | 9.077896 | 88 | 21.022496 |
| 39 | 9.316788 | 89 | 21.261388 |
| 40 | 9.555680 | 90 | 21.500280 |
| 41 | 9.794572 | 91 | 21.739172 |
| 42 | 10.033464 | 92 | 21.978064 |
| 43 | 10.272356 | 93 | 22.216956 |
| 44 | 10.511248 | 94 | 22.455848 |
| 45 | 10.750140 | 95 | 22.694740 |
| 46 | 10.989032 | 96 | 22.933632 |
| 47 | 11.227924 | 97 | 23.172524 |
| 48 | 11.466816 | 98 | 23.411416 |
| 49 | 11.705708 | 99 | 23.650308 |
| 50 | 11.944600 | 100 | 23.889200 |

## POWER
### Horsepower (metric) to Kilowatts

| hp | kW | hp | kW |
|---|---|---|---|
| 1 | .735499 | 51 | 37.510449 |
| 2 | 1.470998 | 52 | 38.245948 |
| 3 | 2.206497 | 53 | 38.981447 |
| 4 | 2.941996 | 54 | 39.716946 |
| 5 | 3.677495 | 55 | 40.452445 |
| 6 | 4.412994 | 56 | 41.187944 |
| 7 | 5.148493 | 57 | 41.923443 |
| 8 | 5.883992 | 58 | 42.658942 |
| 9 | 6.619491 | 59 | 43.394441 |
| 10 | 7.354990 | 60 | 44.129940 |
| 11 | 8.090489 | 61 | 44.865439 |
| 12 | 8.825988 | 62 | 45.600938 |
| 13 | 9.561487 | 63 | 46.336437 |
| 14 | 10.296986 | 64 | 47.071936 |
| 15 | 11.032485 | 65 | 47.807435 |
| 16 | 11.767984 | 66 | 48.542934 |
| 17 | 12.503483 | 67 | 49.278433 |
| 18 | 13.238982 | 68 | 50.013932 |
| 19 | 13.974481 | 69 | 50.749431 |
| 20 | 14.709980 | 70 | 51.484930 |
| 21 | 15.445479 | 71 | 52.220429 |
| 22 | 16.180978 | 72 | 52.955928 |
| 23 | 16.916477 | 73 | 53.691427 |
| 24 | 17.651976 | 74 | 54.426926 |
| 25 | 18.387475 | 75 | 55.162425 |
| 26 | 19.122974 | 76 | 55.897924 |
| 27 | 19.858473 | 77 | 56.633423 |
| 28 | 20.593972 | 78 | 57.368922 |
| 29 | 21.329471 | 79 | 58.104421 |
| 30 | 22.064970 | 80 | 58.839920 |
| 31 | 22.800469 | 81 | 59.575419 |
| 32 | 23.535968 | 82 | 60.310918 |
| 33 | 24.271467 | 83 | 61.046417 |
| 34 | 25.006966 | 84 | 61.781916 |
| 35 | 25.742465 | 85 | 62.517415 |
| 36 | 26.477964 | 86 | 63.252914 |
| 37 | 27.213463 | 87 | 63.988413 |
| 38 | 27.948962 | 88 | 64.723912 |
| 39 | 28.684461 | 89 | 65.459411 |
| 40 | 29.419960 | 90 | 66.194910 |
| 41 | 30.155459 | 91 | 66.930409 |
| 42 | 30.890958 | 92 | 67.665908 |
| 43 | 31.626457 | 93 | 68.401407 |
| 44 | 32.361956 | 94 | 69.136906 |
| 45 | 33.097455 | 95 | 69.872405 |
| 46 | 33.832954 | 96 | 70.607904 |
| 47 | 34.568453 | 97 | 71.343403 |
| 48 | 35.303952 | 98 | 72.078902 |
| 49 | 36.039451 | 99 | 72.814401 |
| 50 | 36.774950 | 100 | 73.549900 |

## POWER
### Kilowatts to Horsepower (metric)

| kW | hp | kW | hp |
|---|---|---|---|
| 1 | 1.359621 | 51 | 69.340671 |
| 2 | 2.719242 | 52 | 70.700292 |
| 3 | 4.078863 | 53 | 72.059913 |
| 4 | 5.438484 | 54 | 73.419534 |
| 5 | 6.798105 | 55 | 74.779155 |
| 6 | 8.157726 | 56 | 76.138776 |
| 7 | 9.517347 | 57 | 77.498397 |
| 8 | 10.876968 | 58 | 78.858018 |
| 9 | 12.236589 | 59 | 80.217639 |
| 10 | 13.596210 | 60 | 81.577260 |
| 11 | 14.955831 | 61 | 82.936881 |
| 12 | 16.315452 | 62 | 84.296502 |
| 13 | 17.675073 | 63 | 85.656123 |
| 14 | 19.034694 | 64 | 87.015744 |
| 15 | 20.394315 | 65 | 88.375365 |
| 16 | 21.753936 | 66 | 89.734986 |
| 17 | 23.113557 | 67 | 91.094607 |
| 18 | 24.473178 | 68 | 92.454228 |
| 19 | 25.832799 | 69 | 93.813849 |
| 20 | 27.192420 | 70 | 95.173470 |
| 21 | 28.552041 | 71 | 96.533091 |
| 22 | 29.911662 | 72 | 97.892712 |
| 23 | 31.271283 | 73 | 99.252333 |
| 24 | 32.630904 | 74 | 100.611954 |
| 25 | 33.990525 | 75 | 101.971575 |
| 26 | 35.350146 | 76 | 103.331196 |
| 27 | 36.709767 | 77 | 104.690817 |
| 28 | 38.069388 | 78 | 106.050438 |
| 29 | 39.429009 | 79 | 107.410059 |
| 30 | 40.788630 | 80 | 108.769680 |
| 31 | 42.148251 | 81 | 110.129301 |
| 32 | 43.507872 | 82 | 111.488922 |
| 33 | 44.867493 | 83 | 112.848543 |
| 34 | 46.227114 | 84 | 114.208164 |
| 35 | 47.586735 | 85 | 115.567785 |
| 36 | 48.946356 | 86 | 116.927406 |
| 37 | 50.305977 | 87 | 118.287027 |
| 38 | 51.665598 | 88 | 119.646648 |
| 39 | 53.025219 | 89 | 121.006269 |
| 40 | 54.384840 | 90 | 122.365890 |
| 41 | 55.744461 | 91 | 123.725511 |
| 42 | 57.104082 | 92 | 125.085132 |
| 43 | 58.463703 | 93 | 126.444753 |
| 44 | 59.823324 | 94 | 127.804374 |
| 45 | 61.182945 | 95 | 129.163995 |
| 46 | 62.542566 | 96 | 130.523616 |
| 47 | 63.902187 | 97 | 131.883237 |
| 48 | 65.261808 | 98 | 133.242858 |
| 49 | 66.621429 | 99 | 134.602479 |
| 50 | 67.981050 | 100 | 135.962100 |

## POWER
### Horsepower (FPS) to Kilowatts

| hp | kW | hp | kW |
|---|---|---|---|
| 1 | .745700 | 51 | 38.030700 |
| 2 | 1.491400 | 52 | 38.776400 |
| 3 | 2.237100 | 53 | 39.522100 |
| 4 | 2.982800 | 54 | 40.267800 |
| 5 | 3.728500 | 55 | 41.013500 |
| 6 | 4.474200 | 56 | 41.759200 |
| 7 | 5.219900 | 57 | 42.504900 |
| 8 | 5.965600 | 58 | 43.250600 |
| 9 | 6.711300 | 59 | 43.996300 |
| 10 | 7.457000 | 60 | 44.742000 |
| 11 | 8.202700 | 61 | 45.487700 |
| 12 | 8.948400 | 62 | 46.233400 |
| 13 | 9.694100 | 63 | 46.979100 |
| 14 | 10.439800 | 64 | 47.724800 |
| 15 | 11.185500 | 65 | 48.470500 |
| 16 | 11.931200 | 66 | 49.216200 |
| 17 | 12.676900 | 67 | 49.961900 |
| 18 | 13.422600 | 68 | 50.707600 |
| 19 | 14.168300 | 69 | 51.453300 |
| 20 | 14.914000 | 70 | 52.199000 |
| 21 | 15.659700 | 71 | 52.944700 |
| 22 | 16.405400 | 72 | 53.690400 |
| 23 | 17.151100 | 73 | 54.436100 |
| 24 | 17.896800 | 74 | 55.181800 |
| 25 | 18.642500 | 75 | 55.927500 |
| 26 | 19.388200 | 76 | 56.673200 |
| 27 | 20.133900 | 77 | 57.418900 |
| 28 | 20.879600 | 78 | 58.164600 |
| 29 | 21.625300 | 79 | 58.910300 |
| 30 | 22.371000 | 80 | 59.656000 |
| 31 | 23.116700 | 81 | 60.401700 |
| 32 | 23.862400 | 82 | 61.147400 |
| 33 | 24.608100 | 83 | 61.893100 |
| 34 | 25.353800 | 84 | 62.638800 |
| 35 | 26.099500 | 85 | 63.384500 |
| 36 | 26.845200 | 86 | 64.130200 |
| 37 | 27.590900 | 87 | 64.875900 |
| 38 | 28.336600 | 88 | 65.621600 |
| 39 | 29.082300 | 89 | 66.367300 |
| 40 | 29.828000 | 90 | 67.113000 |
| 41 | 30.573700 | 91 | 67.858700 |
| 42 | 31.319400 | 92 | 68.604400 |
| 43 | 32.065100 | 93 | 69.350100 |
| 44 | 32.810800 | 94 | 70.095800 |
| 45 | 33.556500 | 95 | 70.841500 |
| 46 | 34.302200 | 96 | 71.587200 |
| 47 | 35.047900 | 97 | 72.332900 |
| 48 | 35.793600 | 98 | 73.078600 |
| 49 | 36.539300 | 99 | 73.824300 |
| 50 | 37.285000 | 100 | 74.570000 |

## POWER
### Kilowatts to Horsepower (FPS)

| kW | hp | kW | hp |
|---|---|---|---|
| 1 | 1.341022 | 51 | 68.392122 |
| 2 | 2.682044 | 52 | 69.733144 |
| 3 | 4.023066 | 53 | 71.074166 |
| 4 | 5.364088 | 54 | 72.415188 |
| 5 | 6.705110 | 55 | 73.756210 |
| 6 | 8.046132 | 56 | 75.097232 |
| 7 | 9.387154 | 57 | 76.438254 |
| 8 | 10.728176 | 58 | 77.779276 |
| 9 | 12.069198 | 59 | 79.120298 |
| 10 | 13.410220 | 60 | 80.461320 |
| 11 | 14.751242 | 61 | 81.802342 |
| 12 | 16.092264 | 62 | 83.143364 |
| 13 | 17.433286 | 63 | 84.484386 |
| 14 | 18.774308 | 64 | 85.825408 |
| 15 | 20.115330 | 65 | 87.166430 |
| 16 | 21.456352 | 66 | 88.507452 |
| 17 | 22.797374 | 67 | 89.848474 |
| 18 | 24.138396 | 68 | 91.189496 |
| 19 | 25.479418 | 69 | 92.530518 |
| 20 | 26.820440 | 70 | 93.871540 |
| 21 | 28.161462 | 71 | 95.212562 |
| 22 | 29.502484 | 72 | 96.553584 |
| 23 | 30.843506 | 73 | 97.894606 |
| 24 | 32.184528 | 74 | 99.235628 |
| 25 | 33.525550 | 75 | 100.576650 |
| 26 | 34.866572 | 76 | 101.917672 |
| 27 | 36.207594 | 77 | 103.258694 |
| 28 | 37.548616 | 78 | 104.599716 |
| 29 | 38.889638 | 79 | 105.940738 |
| 30 | 40.230660 | 80 | 107.281760 |
| 31 | 41.571682 | 81 | 108.622782 |
| 32 | 42.912704 | 82 | 109.963804 |
| 33 | 44.253726 | 83 | 111.304826 |
| 34 | 45.594748 | 84 | 112.645848 |
| 35 | 46.935770 | 85 | 113.986870 |
| 36 | 48.276792 | 86 | 115.327892 |
| 37 | 49.617814 | 87 | 116.668914 |
| 38 | 50.958836 | 88 | 118.009936 |
| 39 | 52.299858 | 89 | 119.350958 |
| 40 | 53.640880 | 90 | 120.691980 |
| 41 | 54.981902 | 91 | 122.033002 |
| 42 | 56.322924 | 92 | 123.374024 |
| 43 | 57.663946 | 93 | 124.715046 |
| 44 | 59.004968 | 94 | 126.056068 |
| 45 | 60.345990 | 95 | 127.397090 |
| 46 | 61.687012 | 96 | 128.738112 |
| 47 | 63.028034 | 97 | 130.079134 |
| 48 | 64.369056 | 98 | 131.420156 |
| 49 | 65.710078 | 99 | 132.761178 |
| 50 | 67.051100 | 100 | 134.102200 |

# POWER

## Horsepower (boiler) to Kilowatts

| hp | kW | hp | kW |
|---|---|---|---|
| 1 | 9.809500 | 51 | 500.284500 |
| 2 | 19.619000 | 52 | 510.094000 |
| 3 | 29.428500 | 53 | 519.903500 |
| 4 | 39.238000 | 54 | 529.713000 |
| 5 | 49.047500 | 55 | 539.522500 |
| 6 | 58.857000 | 56 | 549.332000 |
| 7 | 68.666500 | 57 | 559.141500 |
| 8 | 78.476000 | 58 | 568.951000 |
| 9 | 88.285500 | 59 | 578.760500 |
| 10 | 98.095000 | 60 | 588.570000 |
| 11 | 107.904500 | 61 | 598.379500 |
| 12 | 117.714000 | 62 | 608.189000 |
| 13 | 127.523500 | 63 | 617.998500 |
| 14 | 137.333000 | 64 | 627.808000 |
| 15 | 147.142500 | 65 | 637.617500 |
| 16 | 156.952000 | 66 | 647.427000 |
| 17 | 166.761500 | 67 | 657.236500 |
| 18 | 176.571000 | 68 | 667.046000 |
| 19 | 186.380500 | 69 | 676.855500 |
| 20 | 196.190000 | 70 | 686.665000 |
| 21 | 205.999500 | 71 | 696.474500 |
| 22 | 215.809000 | 72 | 706.284000 |
| 23 | 225.618500 | 73 | 716.093500 |
| 24 | 235.428000 | 74 | 725.903000 |
| 25 | 245.237500 | 75 | 735.712500 |
| 26 | 255.047000 | 76 | 745.522000 |
| 27 | 264.856500 | 77 | 755.331500 |
| 28 | 274.666000 | 78 | 765.141000 |
| 29 | 284.475500 | 79 | 774.950500 |
| 30 | 294.285000 | 80 | 784.760000 |
| 31 | 304.094500 | 81 | 794.569500 |
| 32 | 313.904000 | 82 | 804.379000 |
| 33 | 323.713500 | 83 | 814.188500 |
| 34 | 333.523000 | 84 | 823.998000 |
| 35 | 343.332500 | 85 | 833.807500 |
| 36 | 353.142000 | 86 | 843.617000 |
| 37 | 362.951500 | 87 | 853.426500 |
| 38 | 372.761000 | 88 | 863.236000 |
| 39 | 382.570500 | 89 | 873.045500 |
| 40 | 392.380000 | 90 | 882.855000 |
| 41 | 402.189500 | 91 | 892.664500 |
| 42 | 411.999000 | 92 | 902.474000 |
| 43 | 421.808500 | 93 | 912.283500 |
| 44 | 431.618000 | 94 | 922.093000 |
| 45 | 441.427500 | 95 | 931.902500 |
| 46 | 451.237000 | 96 | 941.712000 |
| 47 | 461.046500 | 97 | 951.521500 |
| 48 | 470.856000 | 98 | 961.331000 |
| 49 | 480.665500 | 99 | 971.140500 |
| 50 | 490.475000 | 100 | 980.950000 |

## POWER
### Kilowatts to Horsepower (boiler)

| kW | hp | kW | hp |
|---|---|---|---|
| 1 | .101942 | 51 | 5.199042 |
| 2 | .203884 | 52 | 5.300984 |
| 3 | .305826 | 53 | 5.402926 |
| 4 | .407768 | 54 | 5.504868 |
| 5 | .509710 | 55 | 5.606810 |
| 6 | .611652 | 56 | 5.708752 |
| 7 | .713594 | 57 | 5.810694 |
| 8 | .815536 | 58 | 5.912636 |
| 9 | .917478 | 59 | 6.014578 |
| 10 | 1.019420 | 60 | 6.116520 |
| 11 | 1.121362 | 61 | 6.218462 |
| 12 | 1.223304 | 62 | 6.320404 |
| 13 | 1.325246 | 63 | 6.422346 |
| 14 | 1.427188 | 64 | 6.524288 |
| 15 | 1.529130 | 65 | 6.626230 |
| 16 | 1.631072 | 66 | 6.728172 |
| 17 | 1.733014 | 67 | 6.830114 |
| 18 | 1.834956 | 68 | 6.932056 |
| 19 | 1.936898 | 69 | 7.033998 |
| 20 | 2.038840 | 70 | 7.135940 |
| 21 | 2.140782 | 71 | 7.237882 |
| 22 | 2.242724 | 72 | 7.339824 |
| 23 | 2.344666 | 73 | 7.441766 |
| 24 | 2.446608 | 74 | 7.543708 |
| 25 | 2.548550 | 75 | 7.645650 |
| 26 | 2.650492 | 76 | 7.747592 |
| 27 | 2.752434 | 77 | 7.849534 |
| 28 | 2.854376 | 78 | 7.951476 |
| 29 | 2.956318 | 79 | 8.053418 |
| 30 | 3.058260 | 80 | 8.155360 |
| 31 | 3.160202 | 81 | 8.257302 |
| 32 | 3.262144 | 82 | 8.359244 |
| 33 | 3.364086 | 83 | 8.461186 |
| 34 | 3.466028 | 84 | 8.563128 |
| 35 | 3.567970 | 85 | 8.665070 |
| 36 | 3.669912 | 86 | 8.767012 |
| 37 | 3.771854 | 87 | 8.868954 |
| 38 | 3.873796 | 88 | 8.970896 |
| 39 | 3.975738 | 89 | 9.072838 |
| 40 | 4.077680 | 90 | 9.174780 |
| 41 | 4.179622 | 91 | 9.276722 |
| 42 | 4.281564 | 92 | 9.378664 |
| 43 | 4.383506 | 93 | 9.480606 |
| 44 | 4.485448 | 94 | 9.582548 |
| 45 | 4.587390 | 95 | 9.684490 |
| 46 | 4.689332 | 96 | 9.786432 |
| 47 | 4.791274 | 97 | 9.888374 |
| 48 | 4.893216 | 98 | 9.990316 |
| 49 | 4.995158 | 99 | 10.092258 |
| 50 | 5.097100 | 100 | 10.194200 |

## POWER
### Horsepower (electric) to Kilowatts

| hp | kW | hp | kW |
|---|---|---|---|
| 1 | .746000 | 51 | 38.046000 |
| 2 | 1.492000 | 52 | 38.792000 |
| 3 | 2.238000 | 53 | 39.538000 |
| 4 | 2.984000 | 54 | 40.284000 |
| 5 | 3.730000 | 55 | 41.030000 |
| 6 | 4.476000 | 56 | 41.776000 |
| 7 | 5.222000 | 57 | 42.522000 |
| 8 | 5.968000 | 58 | 43.268000 |
| 9 | 6.714000 | 59 | 44.014000 |
| 10 | 7.460000 | 60 | 44.760000 |
| 11 | 8.206000 | 61 | 45.506000 |
| 12 | 8.952000 | 62 | 46.252000 |
| 13 | 9.698000 | 63 | 46.998000 |
| 14 | 10.444000 | 64 | 47.744000 |
| 15 | 11.190000 | 65 | 48.490000 |
| 16 | 11.936000 | 66 | 49.236000 |
| 17 | 12.682000 | 67 | 49.982000 |
| 18 | 13.428000 | 68 | 50.728000 |
| 19 | 14.174000 | 69 | 51.474000 |
| 20 | 14.920000 | 70 | 52.220000 |
| 21 | 15.666000 | 71 | 52.966000 |
| 22 | 16.412000 | 72 | 53.712000 |
| 23 | 17.158000 | 73 | 54.458000 |
| 24 | 17.904000 | 74 | 55.204000 |
| 25 | 18.650000 | 75 | 55.950000 |
| 26 | 19.396000 | 76 | 56.696000 |
| 27 | 20.142000 | 77 | 57.442000 |
| 28 | 20.888000 | 78 | 58.188000 |
| 29 | 21.634000 | 79 | 58.934000 |
| 30 | 22.380000 | 80 | 59.680000 |
| 31 | 23.126000 | 81 | 60.426000 |
| 32 | 23.872000 | 82 | 61.172000 |
| 33 | 24.618000 | 83 | 61.918000 |
| 34 | 25.364000 | 84 | 62.664000 |
| 35 | 26.110000 | 85 | 63.410000 |
| 36 | 26.856000 | 86 | 64.156000 |
| 37 | 27.602000 | 87 | 64.902000 |
| 38 | 28.348000 | 88 | 65.648000 |
| 39 | 29.094000 | 89 | 66.394000 |
| 40 | 29.840000 | 90 | 67.140000 |
| 41 | 30.586000 | 91 | 67.886000 |
| 42 | 31.332000 | 92 | 68.632000 |
| 43 | 32.078000 | 93 | 69.378000 |
| 44 | 32.824000 | 94 | 70.124000 |
| 45 | 33.570000 | 95 | 70.870000 |
| 46 | 34.316000 | 96 | 71.616000 |
| 47 | 35.062000 | 97 | 72.362000 |
| 48 | 35.808000 | 98 | 73.108000 |
| 49 | 36.554000 | 99 | 73.854000 |
| 50 | 37.300000 | 100 | 74.600000 |

## POWER
### Kilowatts to Horsepower (electric)

| kW | hp | kW | hp |
|---|---|---|---|
| 1 | 1.340483 | 51 | 68.364633 |
| 2 | 2.680966 | 52 | 69.705116 |
| 3 | 4.021449 | 53 | 71.045599 |
| 4 | 5.361932 | 54 | 72.386082 |
| 5 | 6.702415 | 55 | 73.726565 |
| 6 | 8.042898 | 56 | 75.067048 |
| 7 | 9.383381 | 57 | 76.407531 |
| 8 | 10.723864 | 58 | 77.748014 |
| 9 | 12.064347 | 59 | 79.088497 |
| 10 | 13.404830 | 60 | 80.428980 |
| 11 | 14.745313 | 61 | 81.769463 |
| 12 | 16.085796 | 62 | 83.109946 |
| 13 | 17.426279 | 63 | 84.450429 |
| 14 | 18.766762 | 64 | 85.790912 |
| 15 | 20.107245 | 65 | 87.131395 |
| 16 | 21.447728 | 66 | 88.471878 |
| 17 | 22.788211 | 67 | 89.812361 |
| 18 | 24.128694 | 68 | 91.152844 |
| 19 | 25.469177 | 69 | 92.493327 |
| 20 | 26.809660 | 70 | 93.833810 |
| 21 | 28.150143 | 71 | 95.174293 |
| 22 | 29.490626 | 72 | 96.514776 |
| 23 | 30.831109 | 73 | 97.855259 |
| 24 | 32.171592 | 74 | 99.195742 |
| 25 | 33.512075 | 75 | 100.536225 |
| 26 | 34.852558 | 76 | 101.876708 |
| 27 | 36.193041 | 77 | 103.217191 |
| 28 | 37.533524 | 78 | 104.557674 |
| 29 | 38.874007 | 79 | 105.898157 |
| 30 | 40.214490 | 80 | 107.238640 |
| 31 | 41.554973 | 81 | 108.579123 |
| 32 | 42.895456 | 82 | 109.919606 |
| 33 | 44.235939 | 83 | 111.260089 |
| 34 | 45.576422 | 84 | 112.600572 |
| 35 | 46.916905 | 85 | 113.941055 |
| 36 | 48.257388 | 86 | 115.281538 |
| 37 | 49.597871 | 87 | 116.622021 |
| 38 | 50.938354 | 88 | 117.962504 |
| 39 | 52.278837 | 89 | 119.302987 |
| 40 | 53.619320 | 90 | 120.643470 |
| 41 | 54.959803 | 91 | 121.983953 |
| 42 | 56.300286 | 92 | 123.324436 |
| 43 | 57.640769 | 93 | 124.664919 |
| 44 | 58.981252 | 94 | 126.005402 |
| 45 | 60.321735 | 95 | 127.345885 |
| 46 | 61.662218 | 96 | 128.686368 |
| 47 | 63.002701 | 97 | 130.026851 |
| 48 | 64.343184 | 98 | 131.367334 |
| 49 | 65.683667 | 99 | 132.707817 |
| 50 | 67.024150 | 100 | 134.048300 |

## POWER
### Horsepower (water) to Kilowatts

| hp | kW | hp | kW |
|---|---|---|---|
| 1 | .746043 | 51 | 38.048193 |
| 2 | 1.492086 | 52 | 38.794236 |
| 3 | 2.238129 | 53 | 39.540279 |
| 4 | 2.984172 | 54 | 40.286322 |
| 5 | 3.730215 | 55 | 41.032365 |
| 6 | 4.476258 | 56 | 41.778408 |
| 7 | 5.222301 | 57 | 42.524451 |
| 8 | 5.968344 | 58 | 43.270494 |
| 9 | 6.714387 | 59 | 44.016537 |
| 10 | 7.460430 | 60 | 44.762580 |
| 11 | 8.206473 | 61 | 45.508623 |
| 12 | 8.952516 | 62 | 46.254666 |
| 13 | 9.698559 | 63 | 47.000709 |
| 14 | 10.444602 | 64 | 47.746752 |
| 15 | 11.190645 | 65 | 48.492795 |
| 16 | 11.936688 | 66 | 49.238838 |
| 17 | 12.682731 | 67 | 49.984881 |
| 18 | 13.428774 | 68 | 50.730924 |
| 19 | 14.174817 | 69 | 51.476967 |
| 20 | 14.920860 | 70 | 52.223010 |
| 21 | 15.666903 | 71 | 52.969053 |
| 22 | 16.412946 | 72 | 53.715096 |
| 23 | 17.158989 | 73 | 54.461139 |
| 24 | 17.905032 | 74 | 55.207182 |
| 25 | 18.651075 | 75 | 55.953225 |
| 26 | 19.397118 | 76 | 56.699268 |
| 27 | 20.143161 | 77 | 57.445311 |
| 28 | 20.889204 | 78 | 58.191354 |
| 29 | 21.635247 | 79 | 58.937397 |
| 30 | 22.381290 | 80 | 59.683440 |
| 31 | 23.127333 | 81 | 60.429483 |
| 32 | 23.873376 | 82 | 61.175526 |
| 33 | 24.619419 | 83 | 61.921569 |
| 34 | 25.365462 | 84 | 62.667612 |
| 35 | 26.111505 | 85 | 63.413655 |
| 36 | 26.857548 | 86 | 64.159698 |
| 37 | 27.603591 | 87 | 64.905741 |
| 38 | 28.349634 | 88 | 65.651784 |
| 39 | 29.095677 | 89 | 66.397827 |
| 40 | 29.841720 | 90 | 67.143870 |
| 41 | 30.587763 | 91 | 67.889913 |
| 42 | 31.333806 | 92 | 68.635956 |
| 43 | 32.079849 | 93 | 69.381999 |
| 44 | 32.825892 | 94 | 70.128042 |
| 45 | 33.571935 | 95 | 70.874085 |
| 46 | 34.317978 | 96 | 71.620128 |
| 47 | 35.064021 | 97 | 72.366171 |
| 48 | 35.810064 | 98 | 73.112214 |
| 49 | 36.556107 | 99 | 73.858257 |
| 50 | 37.302150 | 100 | 74.604300 |

## POWER
### Kilowatts to Horsepower (water)

| kW | hp | kW | hp |
|---|---|---|---|
| 1 | 1.340405 | 51 | 68.360655 |
| 2 | 2.680810 | 52 | 69.701060 |
| 3 | 4.021215 | 53 | 71.041465 |
| 4 | 5.361620 | 54 | 72.381870 |
| 5 | 6.702025 | 55 | 73.722275 |
| 6 | 8.042430 | 56 | 75.062680 |
| 7 | 9.382835 | 57 | 76.403085 |
| 8 | 10.723240 | 58 | 77.743490 |
| 9 | 12.063645 | 59 | 79.083895 |
| 10 | 13.404050 | 60 | 80.424300 |
| 11 | 14.744455 | 61 | 81.764705 |
| 12 | 16.084860 | 62 | 83.105110 |
| 13 | 17.425265 | 63 | 84.445515 |
| 14 | 18.765670 | 64 | 85.785920 |
| 15 | 20.106075 | 65 | 87.126325 |
| 16 | 21.446480 | 66 | 88.466730 |
| 17 | 22.786885 | 67 | 89.807135 |
| 18 | 24.127290 | 68 | 91.147540 |
| 19 | 25.467695 | 69 | 92.487945 |
| 20 | 26.808100 | 70 | 93.828350 |
| 21 | 28.148505 | 71 | 95.168755 |
| 22 | 29.488910 | 72 | 96.509160 |
| 23 | 30.829315 | 73 | 97.849565 |
| 24 | 32.169720 | 74 | 99.189970 |
| 25 | 33.510125 | 75 | 100.530375 |
| 26 | 34.850530 | 76 | 101.870780 |
| 27 | 36.190935 | 77 | 103.211185 |
| 28 | 37.531340 | 78 | 104.551590 |
| 29 | 38.871745 | 79 | 105.891995 |
| 30 | 40.212150 | 80 | 107.232400 |
| 31 | 41.552555 | 81 | 108.572805 |
| 32 | 42.892960 | 82 | 109.913210 |
| 33 | 44.233365 | 83 | 111.253615 |
| 34 | 45.573770 | 84 | 112.594020 |
| 35 | 46.914175 | 85 | 113.934425 |
| 36 | 48.254580 | 86 | 115.274830 |
| 37 | 49.594985 | 87 | 116.615235 |
| 38 | 50.935390 | 88 | 117.955640 |
| 39 | 52.275795 | 89 | 119.296045 |
| 40 | 53.616200 | 90 | 120.636450 |
| 41 | 54.956605 | 91 | 121.976855 |
| 42 | 56.297010 | 92 | 123.317260 |
| 43 | 57.637415 | 93 | 124.657665 |
| 44 | 58.977820 | 94 | 125.998070 |
| 45 | 60.318225 | 95 | 127.338475 |
| 46 | 61.658630 | 96 | 128.678880 |
| 47 | 62.999035 | 97 | 130.019285 |
| 48 | 64.339440 | 98 | 131.359690 |
| 49 | 65.679845 | 99 | 132.700095 |
| 50 | 67.020250 | 100 | 134.040500 |

# INDEX

## D

## H



## I



## J

## K



## L

## M

## N



## O



## P



## Q